Marijuana
PEST & DISEASE
CONTROL

Marijuana Pest & Disease Control
Copyright © 2012 Ed Rosenthal

Published by Quick American
A Division of Quick Trading Co.
Oakland, CA

ISBN: 978-0932551047
Printed in Korea
Third Printing

Project Director: Jane Klein
Editors: S. Newhart, Eva Saelens, Grubbycup Stash
Entomology editor: Ken Ahlstrom, PhD
Photo Editior: Hera Lee
Design and Production: Alvaro Villanueva
Illustrations by Steve Buchanan © The Taunton Press, Inc.

FRONT COVER: left image: Ed Rosenthal; right top to bottom: Ed Rosenthal; Alex Wild; Scott Bauer,
USDA Agricultural Research Service; TCurtiss. BACK COVER: left top to bottom: Ed Rosenthal;
Ed Rosenthal; Sukalo. Right top to bottom: Ed Rosenthal; Ed Rosenthal; Clemson University - USDA
Cooperative Extension Slide Series, Bugwood.org. INTERIOR PHOTO CREDITS: page 246.

Library of Congress Cataloging-in-Publication Data

Rosenthal, Ed.
 Marijuana pest and disease control / Ed Rosenthal ; with Kathy Imbriani.
     p.   cm.
 Includes bibliographical references.
 ISBN 978-0-932551-04-7
 1. Marijuana--Diseases and pests--Control. 2. Cannabis--Diseases and pests--Control. I. Imbriani, Kathy. II. Title.
 SB608.C28R67 2012
 633.5'3--dc23
                    2012004251

Marijuana
PEST & DISEASE CONTROL

Ed Rosenthal
with Kathy Imbriani

QUICK AMERICAN PUBLISHING

"Winter grey and falling rain, we'll see summer come again,
Darkness falls and seasons change (gonna happen every time).
Same old friends the wind and rain, summers fade and roses die,
You'll see summer come again, like a song that's born to soar the sky."

Lyrics: Eric Andersen and Bob Weir, "Weather Report Suite Part 1."
Courtesy of the Grateful Dead

ACKNOWLEDGMENTS

I would like to thank and acknowledge the team that helped to put this work out. They are Angela Bacca, Anna Foster, Kathy Imbriani, Jane Klein, Hera Lee, Eva Saelens, Jason Schulz, J.C. Stitch, Alvaro Villanueva, and Shelli Newhart Walker.

I would like to give a special acknowledgement to William Quarles for his information and inspiration. He is a best friend whom I have never met or spoken with.

CONTENTS

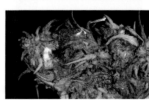

Part 2: CONTROLS **The Problem Solvers**

INTRODUCTION

The purpose of this book is to help you deal with some of the most persistent problems marijuana gardeners face: pests and diseases that thrive on your plants. *Marijuana Pest & Disease Control* provides you with a firm understanding of the problem you are dealing with. Before you begin any competition, you want to know a bit about the other side. You always do better when you have more information, and a better understanding, of what you are up against. Then you can target your efforts and energy on effective solutions rather than striking out indiscriminately.

Arthropods (soft-bodied insects), the main group of pests covered, broke away early from other lines in the animal kingdom. They followed a different course of evolution than the line that led to humans. We live in the same world but sense it in different ways, and use different strategies in facing nature's challenges. Arthropods are part of the texture of the world we live in and adapt to it very successfully. This is why the information in this book is so important. When arthropods' and humans' paths conflict, we have to know how they live. Understanding lifestyles, intimate lives, and goals will help humans stop them from harming our projects, and protect the living organisms under our care and guardianship.

A few years ago the Monterey Bay Aquarium had a show about the deep ocean, described as "Same planet—Different world." But you don't have to put on diving equipment to experience the same phenomenon in the world around us. Just look around any outdoor garden and you will see groups of creatures whose world and sensations of the world are so completely alien from ours that they might as well live on another planet.

Researching this book gave me a new appreciation of all of these organisms and their unique roles in nature. I observed how the will to thrive spurs nature to evolve creatures that may look or seem weird to us, but that have perfect form and function for their lifestyle

We interact with cousins of these herbivore creatures and organisms every day on a very personal level. We are subject to attacks by fungi, bacteria, arthropods that bite, sting, and suck, and on a slightly less personal level, from all types of small mammals. In our everyday world of dealing with these species we tend to depersonalize them and to denigrate their attributes, amazing capabilities, and ability to evolve.

All creatures alive today, including the pests in this book, are survivors, no different than us, struggling to thrive. After 3.5 billion years of evolution, we have all arrived here together. The tiny mite is as much a result of evolution as we are, and it is as unique. We are all the remaining 1 percent or less of all the species trials nature has tossed into the mix.

The reason I bring this up is to warn readers that we should not believe we are superior to these threats. The organisms that cross your path are highly evolved and have adapted many survival techniques. In the war between plant pests, diseases, and humans, the other side has been able to hold its own.

Human-invented pesticides are only about 70 years old. These synthetics have caused untold ecological and health problems. Pests develop resistance to them rather quickly, so there is a constant battle between weapons and armor, or resistance by arthropods. Synthetic pesticides have a mixed record. Many are dangerous to humans, other mammals, birds, reptiles, fish, or crustaceans. Arthropods and diseases evolve and gain resistance to the synthetics alarmingly fast, so more complex chemistry is required. The long-term effects of these chemicals are not completely known, but it is well-established that they can cause damage to both human health and the health of the environment.

Until 1947, with the introduction of DDT, pest control was left to nature. Before then, humans manipulated nature to create natural solutions to problems. For examples, pyrethrum and cinnamon are natural plant-derived insecticides that have been used for hundreds or thousands of years.

Plants have been in a battle with pests for millions of years. Unlike animals, they are rooted and cannot move to escape threats. Instead, plants and fungi have developed an incredible array of chemical weapons to battle predators and diseases. We use this natural chemistry, along with microbial plant allies and other creatures that protect plants by feasting on herbivores, to protect plants from predators and diseases.

Using natural plant chemistry and other natural means to eliminate problems is tried and true. As I mentioned earlier, natural pesticide chemistry has been used by plants with no help from humankind, for millions of years. During this time, the plants honed the efficacy of these protectants, so they work consistently without pests developing resistance.

The solutions in this book are friendly to humans, pets, and the planet, and all are extremely effective. This book shows you how to protect your garden safely.

> Rule No. 1 for foods, cosmetics, fertilizers, and pesticides: If you can't pronounce the ingredients on the label, don't use it.

PART 1:
Pests & Diseases

HOW TO USE THIS BOOK

This book is split into two parts. Part 1 is divided into chapters, with a chapter dedicated to each of the pests and diseases that can afflict cannabis. These chapters include a look at how the pest lives, what it eats, how it reproduces, and the effect of the pest or disease on the plant. Each chapter has descriptions including photos and drawings so you can identify both the symptoms and perpetrator. This section also discusses solutions that are specific to the pest or disease discussed in the chapter.

Part 2 of the book is Controls—The Problem Solvers. This part consists of four sections, each a different strategic approach. For many problems redundancy—using more than one solution—is good because it creates multiple modes of action: "If the thunder don't get you, then the lightning will."

Section 1 discusses physical barriers—how to keep the problems out of the garden and away from your plants. If the pests can't get to the plants, then you don't have to get them out.

Section 2 is devoted to pesticides and fungicides. Each control is described in detail including its benefits, ease of use, efficacy, and contraindications. Its compatibility with other solutions is also discussed. Some brand names of control products are listed but this is a partial list. Mention in this book is not an endorsement of a particular brand, although recognition that this brand is safe to use on plants that will be ingested in some form.

Section 3 is a comprehensive study of biological controls. Biological controls are living things that can prevent other living things from harming your garden. Each control is examined in detail, spanning from carnivorous insects to fungi and bacteria. These living organisms and the products they produce can be introduced into a garden without concern about chemical exposure, and most can be used anytime during the life cycle of the plant.

Section 4 discusses strategies specific to outdoor gardening. Integral topics are covered, such as companion planting, and using animals to fortify defenses against pests.

There is no single correct way to use this book. If the reader has time, I would advise reading it cover to cover. Every marijuana grower, no matter how experienced, will learn a vast amount of information relevant to their own garden from a thorough reading of this book. To paraphrase the war criminal (and former Secretary of Defense) Donald Rumsfeld, there are things you know that you don't know and things you don't know that you don't know. This book will answer many of those "known unknowns" and "unknown unknowns." I certainly learned a lot while writing it.

Another way to use this book is to use it as a reference guide. Let's say you notice a plant problem. Identify the offending pest or disease by looking through Part 1. Find the chapter that covers whatever it is that is plaguing your garden. Then read the chapter devoted to the pest. This will give you an understanding of the threat to the garden, the plants, and the buds. With this knowledge you can make intelligent, informed decisions about how to combat the threat to your efforts.

Each chapter ends with a summary of solutions for that particular pest or disease. You should then turn to Part 2, where you can find more in-depth coverage of the solutions that are available. With this information you can choose the best method to bring your garden to health and vitality, so the vigorous plants produce a high yield of potent buds.

Preventive Maintenance: The Basics

The best first step any gardener can take in terms of pest control is to make pests unwelcome in the garden. How do you do this? Good, clean growing practices go a long way toward helping to avoid pest infestations. Aside from basic cleanliness, there are four main factors in preventive maintenance, which correspond to the garden's main elements: air, water, soil, and plants. The following preventive maintenance practices should be implemented in gardens to minimize opportunities for pests to inhabit the garden.

BASIC CLEANLINESS & SANITATION

Clean With Hydrogen Peroxide

Use hydrogen peroxide (H_2O_2) to sterilize greenhouse benches, pots, tools, and any other equipment used in production. Sterilize pots and trays before reusing them to prevent disease development. Do not allow tools used in sterile soil in indoor plantings to touch the ground or unclean surfaces. Wash down greenhouse benches and side walls after production and allow them to dry completely before introducing new plants to the space.

Planting Media

When growing indoors, it is important to ensure that planting mix used for seeds or starts is pasteurized, sterilized, or inert. Some spores from fungal infections are short-lived, but present a danger because they are airborne. Other spores, called cleistothecia, are more long lived. These spores are in a closed husk, protecting them through drought or winter, to be released only after the shell disintegrates. To be sure that all spores are eliminated and the space is free from infection, spray or wash with a disinfectant such as hydrogen peroxide, or a strong household cleaner.

Foggers, such as this model from Sanitize Systems, are excellent tools for sanitizing rooms and spraying plants because of their thorough coverage.

AIR
Temperature & Humidity
High humidity and low temperatures encourage some plant diseases. Maintain optimum temperature range—70 to 90 °F and humidity of 40 to 50 percent. Increase air circulation between plants to control the humidity and to prevent an ideal environment for some insects. Humidity increases as plants get larger and the canopy closes. Proper ventilation facilitates control of temperature and humidity levels. When the temperature differences between daytime and nighttime are lessened in greenhouses, formation of dew reduces, which in turn reduces humidity and the standing droplets of water or puddles caused by dew condensation.

Proper Plant Spacing
Provide good air circulation for both indoor and outdoor plantings by leaving space between plants. This allows moisture to dry more quickly, eliminating an environmental factor that encourages fungi to form. In sea-of-green gardens, make sure the breeze is strong enough to penetrate the canopy.

Insecticidal Light
UVC light kills fungal spores, bacteria, and insects that come in contact with it. These lights are often used in restaurants. The fixtures enclose the light. A fan draws air through the lights. These lights are very useful in closed rooms, where the air is not exchanged, and for ventilation systems, where incoming air is treated.

Carbon Dioxide (CO_2) Enrichment
Carbon dioxide is used by plants as an ingredient for photosynthesis. Plants combine CO_2 with water, using light as energy. The result is sugar, which is used to supply the plant with building material and energy. The ambient CO_2 level in air is 392 parts-per-million. Plants photosynthesize faster in the presence of CO_2-enriched air. Bright rooms can use up to 1,300 PPM.

WATER
Use A Clean Water Source
Be sure to use a fungus-free source of irrigation water. Many fungal spores and bacteria are water-borne. Infected water can be sterilized using 1 percent hydrogen peroxide or by passing it through a UVC light. UVC sterilizers are sold in kits designed to help maintain ponds.

Eliminate any leaks in the garden space or watering system and make sure drainage is complete. A dripping hose, a puddle, or constantly moist area leads to algae growth, supports fungus growth, and attracts fungus gnats.

Reverse osmosis is a process that uses pressure to push water molecules through a porous membrane. Pathogens, dissolved solids, and harmful chemicals remain on one side of the membrane, but pure water passes through.

Avoid Overwatering
Careful watering is a great start for preventive pest control because it is the key to avoiding the growth of fungi and molds. Most fungi responsible for damping off thrive in overly wet environments. Avoid overwatering seedling trays and ensure that the planting mix drains well in larger containers. Plant seeds ⅛ to ¼ inch deep in a lightweight, well-draining soil mix. Another method is to lay the seeds on top of the plant-

ing mix and then cover lightly with moistened peat moss, vermiculite, or fine sand. Seeds planted deeper than ¼ inch are more vulnerable to damping off. Include vermiculite and or perlite to assure good drainage. In addition, allow the soil to dry slightly between waterings, but not to the point of wilting. Avoid heavy soils, since they encourage damping off organisms to flourish. If you must plant in heavy soils, such as clay and silt soils, amend with mulch, sand, perlite, or sterile compost. These amendments aid aeration and drainage.

Reverse-osmosis systems from GrowoniX eliminate excess dissolved solids and other impurities from the water.

Avoid Drought

Not only too much water will result in fungal development. Drought to the point of wilt weakens plants, making them more vulnerable to fungal infections.

Both overwatering and too-dry conditions are bad for roots. In order to thrive, roots need a constant supply of both oxygen and water. By maintaining moist but not wet conditions, the plants will do well.

Balance the pH: pH Up

Fungi thrive in pH environments lower than 7, which is acidic. They prefer an environment of 4.5 to 6. Raising the pH of leaf surfaces is a simple and inexpensive way to battle an infection or prevent one. The product pH Up is a pH adjustor used in indoors gardens, aquariums, and hydroponic systems. The active ingredient is usually potassium hydroxide (KOH) or potash (K_2CO_3). When using pH Up, make a solution with a pH of 8 and spray the foliage to create an inhospitable environment for fungi.

SOIL
Use Sterile Soil

Use a non-soil planting mix, or pasteurized or sterile potting soil to prevent damping off in seedling trays. Indoors, use planting mix. Make sure the bags have no tears or rips. Uninfected used soil and planting mixes contain beneficial organisms that live in a symbiotic relationship with roots. When new roots grow, the micronutrients are there to greet them. Avoid planting in areas previously infested with *Rhizoctonia*. Sclerotia and mycelia may persist in soil for up to six years.

Optimize Soil Structure & Nutrition

Balance soil nutrients carefully. Avoid excess nitrogen by using a balanced fertilizer. Too much nitrogen makes plants vulnerable to infection and attracts leaf-eating insects. Too much phosphorus used to promote flowering also predisposes plants to invasion from disease organisms and insects.

Mixing worm castings, compost, compost tea, and other organic matter encourages the growth of natural biocontrol organisms. Always use well-composted material. Poorly processed compost may contain pests, seeds, and diseases that were not killed.

Avoid Early Planting

When planting outdoors, avoid planting too early in the spring. Start plants indoors and plant outdoors after the soil has warmed. Most damping off fungi thrive in cool, moist conditions.

Avoid Excess Nitrogen

Excess nitrogen encourages the growth of *Fusarium*. Increase the potassium and the phosphorus in the soil, and keep the nitrogen in check.

Use Compost & Compost Tea

Compost is the dark, rich compound produced from decayed leaves, grass, and garden waste that most good gardeners keep in ready supply. It is rich in nutrients, micronutrients, and beneficial microbes that are barriers to infection. When compost is added to a grow area, it aerates the soil and improves soil structure. Compost tea can be applied by adding it to the soil, by using a foliar spray, or by adding it to hydroponic systems. It can be bought fresh-brewed from some hydroponics shops or purchased as a do-it-yourself kit under names such as Bountea and Vermicorp.

RECIPE FOR MAKING COMPOST TEA

To brew tea from your own compost, you will need:

> *1-gallon bucket (buckets larger than 1 gallon will need additional bubble stones for aeration)*
> *Aquarium pump with hose and bubble stone or strip bubbler attached*
> *Nylon stocking or cheesecloth bag*
> *Enough compost to fill a bucket about one-quarter full*

Fill the stocking or cheesecloth bag with compost. Tie the end of the bag to the bucket handle to suspend it down into the bucket. Fill the bucket with water, put in the bubble stones and hoses down in the bottom. Molasses is often added now to feed the microorganisms. Turn on the pump and allow the water to bubble for one to three days. Turn off the pump and let the tea settle. Strain the liquid through cheesecloth or other strainer and apply full strength as a foliar sprayer or add it to the irrigation water. Spread the leftover compost around plants.

A word or two of caution: After aeration, the liquid in the bucket should be deep brown. It should not have an unpleasant, sour, or acrid smell. If it does smell sour, the mixture was not properly aerated and anaerobic respiration has set in. Unfortunately, if this is the case, you should not use it. You may be able to return it to health by aerating it. Compost tea should be used within an hour or two of preparation because it turns anaerobic quickly, usually within a few hours.

Deep Plow
Outdoor Gardens

For outdoor gardens, tilling the soil or raised bed deeply right after harvest exposes any overwintering pests—adults, larvae, pupae, or eggs. Instead of becoming next year's pests, these insects are weathered by the elements and become a snack for birds and other insect predators. Deep plowing also helps prevent plant diseases. It exposes overwintering spores and sclerotia (which are hardened fungal mycelia) or buries them so deeply that they cannot germinate. Deep tillage also reduces compaction of the soil and helps combat root rot

PLANTS
Quarantine New Plants

Always inspect and quarantine new plants for fungal or insect infestations before introducing them into the general population. Keep them in an isolated location or a separate room until you have established that they are pest-free, which takes about 10 days. Taking precautions to confirm that plants do not have any hitchhikers aboard reduces the workload on pest management later on.

Avoid Planting
Infected Seed

Avoid using seed harvested from female plants infected with disease, especially those infested with gray mold. *Botrytis cinerea* is seed-borne as are many *Fusarium* species. Dust the seeds with fungicide or soak them in hydrogen peroxide or compost tea to eliminate or reduce the chances of propagating the fungus, or simply use other seeds that have not been infected.

Remove Pruned or
Dead Plant Material

Prune out and remove any dead or diseased foliage or branches from the grow space, greenhouse, or outdoor garden. Outdoors, keep the crop weeded and avoid tall, thick weedy areas nearby. Dispose of any crop residues off site.

PREPPING THE OUTDOORS

Indoor gardens are controlled environments, but the outdoors is an open system. This has its pluses and minuses. Outdoor gardens get the benefits of a full ecosystem, where natural predators can help keep insects in check. However, in the outdoors, there are large mammalian pests, such as gophers, rats, and deer that no indoor gardener will ever need worry about. Here are some tips for garden planning outdoors that may help avoid pest problems.

Location, Location, Location

Do your research on locating the outdoor garden. If there are paths you know deer follow through your land, do not plot the garden to be a snack spot along their sojourn. Select a location where fencing presents the least difficulties should it be necessary.

Eliminate Ready-Made Banquets

Do not place bird feeders, compost, or other potential outdoor pest buffets near the garden location or vice versa. If there are outdoor dog pens or other outdoor animals or pets, keep their food bowls clean. Avoid planting attractant plants—especially if you know what type of mammals frequent your yard or farm. Don't supply their favorite snack foods in the form of garden bushes or plants nearby to your garden where there might be some carryover grazing.

Eliminate Ready-Made Shelter

Keep wood piles and debris orderly and off the ground or out of the garden's vicinity. Aside from companion plants, keep the area around the garden trimmed close. There should be no bushes, tall grasses, thick vines, or other convenient nesting or hiding places around the garden.

Indoor Garden Precautions

Never visit your indoor garden after spending time outside. First wash and change into fresh or garden-space clothing. This goes for all workers and visitors as well.

Don't allow pets, especially indoor-outdoor pets, into the garden space.

If the space is located in a shed or near a garden or other vegetation, develop a clear path so you don't come in contact with possibly contaminated plants when walking to the space.

Place a mat in a tray and keep it wet with salt water. Step both shoes onto the mat before entering the space. This kills any shoe hitchhiker that may be traveling with you.

NUTRIENTS: N-P-K

Proper nutrition is a key component to plant health. Strong vigorous plants are much more resistant to problems with insects, fungi, and disease. A plant weakened by malnutrition is an open invitation to the unwelcome guests discussed in the later sections of this book.

Nitrogen (N)

Nitrogen (N) deficiency is the most common nutrient deficiency in cannabis.

A slight deficiency causes lower leaves to turn pale green. As the deficiency continues, lower leaves yellow and die from the leaf tips inward as the mobile element migrates from lower leaves to support new growth. The deficiency travels up the plant until only the new growth is green, leaving the lowest leaves to yellow and wither. At the same time growth slows and the stems and petioles turn a red/purple tinge.

Too much nitrogen causes a lush dark green growth that attracts pests and pathogens. The

Nitrogen (N) deficiency: Lower leaves yellow as N moves to new growth.

stalks become brittle and are liable to break from lack of flexibility.

Plants use nitrogen to produce chlorophyll so it is essential to photosynthesis. It is a component of amino acids, and leaf tissue; without it, growth quickly stops.

Water-soluble nitrogen (especially nitrates, NO_3) is available to the roots and quickly absorbed. Insoluble nitrogen (such as urea) needs to be broken down by microbes in the soil or planting mix before the roots can absorb it. After fertilization, nitrogen-deficient plants absorb N as soon as it is available and start to change from pale to a healthy-looking kelly green.

Nitrogen is the first number of the three-number set found on all fertilizer packages, which list N-P-K, always in that order. Any water-soluble fertilizer much higher in N than P and K can be used to solve N deficiencies very quickly. Most hydro vegetative formulas fall into this category.

Without high amounts of nitrogen, especially during the vegetative growth stage, the plant's yield is greatly reduced. N issues happen throughout the entire growth cycle.

Tapering off the use of nitrogen near the flowering period promotes flowering rather than vegetative growth. However, a small amount of N is always necessary.

Phosphorus (P)

Phosphorus deficiency results in slow growth. Plants become stunted, with small leaves. Older leaves are affected first. They turn dark green and develop dull blue or purple hues. The edges of the leaves turn tan or brown and curl downward as the deficiency continues. The lower leaves turn yellow and die.

PHOSPHORUS DEFICIENCY

Lower leaves turn yellow and die.

Leaf necrosis.

New growth is stunted.

Too much phosphorus tends to lock out zinc, which results in spotting and bleached spots (chlorosis) between the leaf veins and twisted new growth. Plants use high amounts of P during the first stages of growth, when it helps seedlings to germinate, aids in root and stem growth, and influences plant vigor. During flowering it promotes rapid growth. Inadequate amounts result in lower yields.

Phosphorus is the second of the three-number ratio listed on fertilizer packages. Water-soluble fertilizers containing high phosphorus fix the deficiency. Bloom fertilizers are high in phosphorus formulas. High-P guano also provides readily available phosphorus. Rock phosphate and greensand are also high in P but they release it gradually.

Potassium (K)

Potassium (K) deficiency occurs occasionally in both planting mediums and outdoors in soil, but rarely in hydroponics. Many organic fertilizers such as guano, fish emulsion, alfalfa, cottonseed, blood meals, and many animal manures contain minor amounts of potassium relative to nitrogen and phosphorous.

Too little potassium causes plants to look vigorous, even taller than the rest of the population, but the tips and edges of their bottom leaves die or turn tan or brown and develop necrotic spots.

As the deficiency gets more severe the leaves develop chlorotic spots. Mottled patches of red and yellow appear between the veins, which remain green, accompanied by red stems and petioles. More severe deficiencies result in slower growth, especially when plants are in the vegetative stage. Severe potassium shortages cause leaves to grow smaller than usual.

Too much potassium causes fan leaves to show a light to dark yellow or white color between the veins.

Potassium is necessary for all activities having to do with water transportation, as well as all stages of growth; it's especially important in the development of buds. K aids in creating sturdy and thick stems, disease resistance, water respiration, and photosynthesis.

Water-soluble fertilizers containing high potassium fix the deficiency. Liquefied kelp, bloom fertilizers, and wood ash are commonly used and work quickly to correct K deficiencies. So do po-

Potassium (K) deficiency: Chlorotic and necrotic areas on leaf edges.

tassium bicarbonate ($KHCO_3$), potassium sulfate (K_2SO_4), and potassium dihydrogen phosphate (KH_2PO_4). Potassium silicate (K_2SiO_3) can be used to supply Si and has 3 percent K in it. Granite dust and greensand take more time to get to the plant and are not usually used to correct deficiencies, but to prevent them.

SELECTING APPROPRIATE SOLUTIONS

One rule of thumb for controlling pests is to begin with the least invasive method. This book uses an approach that introduces the fewest chemicals or substances into the garden.

When dealing with garden pest control management, you become a diagnostician, your garden's doctor. You may not want to resort immediately to surgery, or heavy drugs with many potential side effects, when a problem may be solved by much simpler means. You would probably also prefer that whatever problem you are experiencing be solved rather than just having the symptoms managed. Similarly in the garden, it is better to start with the simplest, most cost-effective solutions that may have the least impact on the garden and its overall environment.

Likewise, if you address the root cause rather than simply blasting with pesticides, you may be not only ridding yourself of a pest but also improving the garden's health and its eventual harvest.

1. Do not use controls labeled only for ornamentals or turf on any plant you will later ingest or inhale.

2. Be aware of potential allergens. Some solutions come with allergy warnings: *Beauvaria bassiana*, pyrethrum. The reactions to these may be serious. Be aware of this possibility and make sure anyone who will interact with the garden when these solutions are in use has checked to be sure they are not allergic.

3. Check that the solution will not negatively affect other species in your environment. While some solutions aren't dangerous for humans, they may be bad for pets. Others are harmless to mammals, but may endanger bird or sea life. These impacts on the environment should be taken into account when selecting appropriate measures.

Ants

At a glance

Symptoms

Ants visibly crawling up stems, increase in aphid population (which ants domesticate and herd), stickiness from increased "honeydew" production by aphids, signs of cut leaves or stem damage from insect activity, visibly weakened plants caused by insect damage.

Ants tending aphids.

Ants seem commonplace, and that's because they are, perhaps more than you ever realized. These little creatures are the most numerous animals on the planet, comprising 10 percent of animal biomass. Even though the typical worker ant only weighs between 1 and 5 milligrams, when taken collectively, the global ant population equals the weight of the entire human population. Ants have colonized every continent except Antarctica, and are native to all but a few of the more remote islands, such as Hawaii, Polynesia, and Greenland.

Don't let their ordinariness lull you into giving them a pass, especially in an indoor garden. They are not exempt from a zero-tolerance policy on pests. If you see only a few of them, rest assured that there are more where those came from.

These social insects never travel alone for long. You've likely spotted scouts, who are looking for new sources of food and shelter. If they find something good they will report back to the community and there will soon be a steady stream of pedestrians on a mission to relocate in your garden. Whatever their goals, you can be fairly sure that they conflict with yours.

There are estimated to be 22,000 species of ants, of which 12,000 species have been identified. Ants are members of the insect family Formicidae, a Latin derivation of the word "ant." Ants belong to the same taxonomic order, Hymenoptera, as bees, wasps, and sawflies. While we think of ants as crawling insects, they have wings for the brief period during which they mate and disperse, just like their bee, wasp, and sawfly cousins.

Ants can create garden problems in several ways, and whether they do or not depends in part on the species. Some ant species, such as the Argentine, odorous house, and velvety tree ants, are "honeydew" seekers. They herd aphids and other bugs that exude the sugary plant juice concentrate that the ants consume. Other ant species cut leaves to earn a living. They can devastate a plant in less than a day.

Worker field ants exchanging food.

Leafcutter ants farm a domesticated fungus that they eat. They use the cut leaves as the medium for their fungus crop. Fire ants pose another set of problems. They are very territorial, have a formidable sting, attack humans who disturb their nest and other vertebrates and invertebrates, and eat young plants, so they must be eliminated from land where marijuana is grown.

Regardless of type, when ants make their way into an indoor garden and nest in plant containers, they disturb plant roots and can create conditions leading to plant failure. Even if they do not engage in any of these destructive behaviors, ants are still a potential disease vector. An individual ant can live about two months and in that time it traverses many diverse conditions where it may pick up pathogens and carry them back to the nest. Ants can introduce both fungal and bacterial pathogens. For all of these reasons, it is important to keep ants out of your garden.

ANT ANATOMY

Ants range in size from 1 millimeter to more than 40 millimeters, depending on the species. Their bodies have three main segments: a head, a thorax (middle segment), and a more bulbous-shaped gaster (hind segment). Most people feel completely confident in their ability to identify ants. The ant's distinctive body shape, which has a very small, constricted "waist" or petiole that connects the thorax and gaster segments together, is a characteristic it shares with bees and wasps. The petiole consists of either one or two distinct segments. However, ants are sometimes confused with termites, especially during the spring when the winged females are leaving the nest. Here are some differences.

The ant exoskeleton varies in color from tan to black to reddish brown. They have six very strong three-jointed legs that allow them to scurry away quickly when threatened. They also have distinct "elbowed" antennae. The antennae ends are covered in hairs, which serve as the ants' main tactile organs. Other parts of antennae serve as the sense organs that we associate with our noses and tongues. One reason why the olfactory sense is important to ants is that ant colonies have different scents, and ants detect intruders by their signature smell.

Because different species of ants and ants within a colony are adapted to specialized tasks, their head shapes and structures vary greatly. One way that ants differ from one another is the complexity of their eyes. Most ants have two compound eyes—eyes with multiple lenses—on the sides of their heads. The number of individual lenses varies considerably by type of ant, with a few having no eyes at all. Workers may only have one lens or very few, while some

Ants	Termites	
Two pairs of wings.	Wings all the same size.	 termite ant
Veins in wings easily seen.	Veins present but not noticeable and often reduced to veinlike wrinkles.	
Wings do not break off easily.	Wings break off easily. Broken wings are plentiful in swarming area.	
Antennae are elbowed.	Antennae are straight.	
Front pair of wings extend just past end of body.	Both pairs of wings twice as long as body.	
Bodies have very constricted waist.	Bodies straight without waist.	

specialized worker ants possess as many as 1,200 lenses per eye. Generally speaking, compound eyes detect movement well, but don't have very good resolution.

Ants also have a triangle of "simple eyes" on the tops of their heads that are believed to be an older evolutionary development. The simple eyes provide information about movement in the environment.

Ants' most diversified feature is their mouth. Ants are known for their very strong jaws, which are called mandibles. These mandibles can snap shut at incredible speeds and have adapted for the ants' lifestyles. They are used for a wide variety of purposes including digging nests, weaving, capturing and carrying prey, defense, and carrying food or droplets of liquid.

Adult ants cannot swallow solid food. Instead they swallow the juices that they squeeze out of their food. They have two stomachs—one to hold food for themselves and a second "social stomach" used to collect food they cannot digest. They regurgitate it and feed it to the larvae, which have the ability to digest it. Once they process the food, the larvae share the digested food with the adult workers. This is important because only about 10 percent of ants in stationary colonies are foragers.

ANT SOCIETIES

Ants emerged contemporaneously with the dinosaurs. Because they have such large and complexly coordinated societies, we've been tempted to compare them with our own society. However, their intense lockstep loyalty to their colony also suggests that a better metaphor might be a Borg-like super-organism, sharing one consciousness and operating as a single body rather than as individuals. Either way, the colony is the true unit of meaning for this highly social insect. While we might find the prospect of living in such a regimented society undesirable, we remain fascinated with their highly efficient social organization. Ants, on the other hand, are oblivious to us. They possess sensory organs that only allow them to perceive the few centimeters around them. As far as they are concerned, this world belongs to them.

We commonly think of ants' residential arrangements as taking the form of anthills, but ant colonies have more diverse architecture. Some create nests underground or under rocks, others inhabit logs, trees, or forest-floor debris. Most species of ants have at least one queen. The rest of the ant society is non-reproductive workers. Workers are sterile females and they do all the housework. Their job is to build, repair, and defend the nest, and to provide food for the queen and the other adult ants in the colony. Reproductive ants include future queens which have wings, but lose them after mating. They become queens after mating with reproductive males, called drones. Drones have wings they retain until death. Their sole purpose in life is to mate with unfertilized reproductive females, after which they die.

The queen, the largest member of the colony, is a reproductive female who has mated and lost her wings. Once the new queen has mated, her job as a new leader is to leave her old home and establish her own colony. This is called "budding." She finds a good location and digs herself a nest. Then, she lays lots of eggs. When they hatch into white, legless larvae, she feeds this first brood with her own metabolized wing muscles and stored fat until they pupate.

The pupae transform into sterile female workers that dig their way out of the nest and set out to find food for Mom, for themselves, and for the next brood that Mom is laying while they hunt.

Once the colony's first generation is established, the queen's main job is to lay as many eggs as possible. All other tasks are delegated to the colony's members. As the ant population

Formica fusco (California native ant) queen and worker in the nest.

increases, the colony enlarges the nest by adding chambers and galleries. After a few years, the colony produces winged females and males that leave to establish their own colonies.

During its life, the ant undergoes a complete metamorphosis, starting as an egg, progressing through larva and pupa, until it finally reaches its adult stage, a process that takes eight to 12 weeks. Ant eggs vary in size by species. They can be almost as large as the workers themselves, or nearly microscopic. The sex of each ant is determined by whether the egg was fertilized or not.

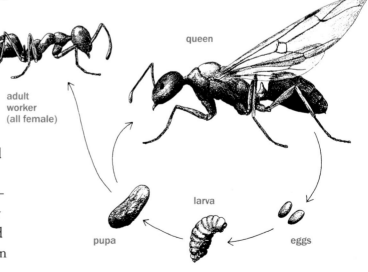

adult worker (all female)

queen

larva

pupa

eggs

Similarities Between Ants and Humans

1) Both live in complex social communities.

2) The communities are cooperative.

3) Both have individuals who will sacrifice themselves for the benefit of the community.

4) Both herd other organisms.

5) Both farm.

6) Both hunt in groups.

7) Both have workers to give the young extended care.

8) Both practice territoriality.

9) Both have forms of slavery.

10) Both have wars between tribes.

11) Both have colonized territories far beyond their native lands.

12) Both have the biggest brains in proportion to body for their phyla.

13) Both have complex modes of communication.

14) Both teach tasks to young members. (Teaching entails performing tasks slowly or in an exaggerated style and paying attention to the student's learning pace.)

15) Both differentiate jobs and assign tasks to specific groups or classes.

16) Individuals change tasks as they mature.

17) They are both capable of complex architecture for homes and workplaces.

18) They both use tools.

19) They both air condition their homes.

20) Each has only a small proportion of members who produce or obtain food for the entire group.

21) Individual ants cannot survive alone. They need the community.

All fertilized eggs become female; unfertilized eggs become the males. Ant eggs hatch into white, legless larvae that are fed regurgitated food by the workers. At night, workers move eggs and larvae deeper into the nest to protect them from the cold. In the morning, they move them closer to the surface of the nest so they can receive the warmth of the sun. By the time the larvae pupate, they resemble adults except they are soft, colorless, and immobile. They emerge from the pupal case as adults. Colonies produce worker ants assigned to specific duties such as defense, larvae nursing, and sanitation. Sanitation workers remove rubbish and waste from the nest to an outside rubbish pile.

ANTS IN THE GREAT OUTDOORS

Ants are part of the natural ecology. There are few places in the world where you won't find them, so you can expect them to be in or around your garden. Your main concern is that they are herding aphids or in the case of leaf-cutter ants, claiming your plants for food. Check to see if ants are going up the stem in a steady stream. This is an indication that you have a problem. In outdoor gardens, conduct at least a little private detective work on the ant world. Where are these ants going, and what are they doing when they get there?

If ants in an outdoor garden aren't climbing up the plant stalks, don't go to war with them. They serve many ecological roles and some of them are helping you. Most of the species that you are likely to encounter are carnivores—insect eaters—and they are keeping the insect population under control. They gang up on caterpillars, grubs, and anything else they can catch.

A true ant story: A gardener had several bushy

12-foot plants that were in the middle of their flowering stage. He noticed ants climbing up the stem of one plant, but could see no damage from the ground. Using a ladder to investigate the phenomenon, he discovered that the ants were butchering and removing dead flesh left over from a small avian predator's meal. This was an unusual situation, but it is an indication of this insect's relentless opportunism, as well as an illustration of one of its roles in the world's ecology.

You should keep an eye out for the ranchers and the cutters. Aphid-herding ants do exactly that—they herd aphids into flocks, sometimes biting off their wings to render them domesticated. Then they wait for the aphids to suck out the plant's juices, and transform it into "honeydew," a sugary substance on which ants love to feed. They milk the dumb suckers for the honeydew they secrete. As the number of aphids increases, which they will at a logarithmic pace, essential life juices will be drained from the plant until it withers from exhaustion.

WHEN ANTS MOVE INDOORS

An indoor garden is a great find for any ant colony that is looking to relocate. If ants gave MVP awards, one would certainly go to the scout that makes such a great discovery. They've discovered the ant promised land: food and shelter without the vagaries of nature's weather. What could make for a cozier nest than planting containers? It is a pristine location, uncolonized by other insects, with nice moist soil. If they were homebuyers, the ants would say, "We'll take it!" What makes the space most convenient is the commute to work. The directions: Go up through the porthole and you'll smell the trail. It leads right up the stem of the plant to the leaf section.

You'll notice the herds of aphids there. Just start milking and bring the honeydew home to the thousands of larvae waiting to be fed.

When an indoor garden suffers from an ant infestation, it may be an indication that the garden needs a hygiene upgrade. Are the ants after something other than the plants and planting medium? Is there human food lying around or something else that the ants might consider eating? The first thing to do before going after the ants is to tidy up your space, with an eye toward eliminating any messes that might be attractive to these pests.

GARDEN-VARIETY ANTS

There are so many species of ants that it is hard to tell one type from another. Luckily, an exact identification is not necessary. Ants vary in prevalence across the globe, but certain types are very common. Argentine ants are found in their original South America, as well as Europe, South Africa, and North America, where they are more of a nuisance on the West Coast than the East Coast. There are several hundred species of stinging ants, but the most common bother to gardeners is fire ants. Two species of fire ants were introduced to the United States. The European fire ant is native to Europe and North Asia, but was first reported in Massachusetts in 1908. A more serious problem emerged from the red imported fire ant, which hitched a ride to the United States from South America during the 1930s. The red imported fire ants are a larger problem in southern U.S. states. Most areas of North America have sugar ants, some examples of which are ghost ants, white-footed ants, odorous house ants, crazy ants, and pharaoh ants. These, along with the Argentine ants, like the sweet stuff.

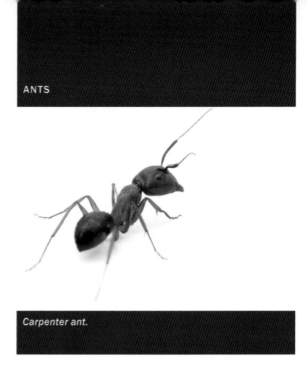

ANTS

Carpenter ants are also common and are named because they nest in wood and can ruin wood structures. Despite rumors to the contrary, they do not actually eat the wood. They feed mainly on a diet of protein and sugars, which outdoors consists of dead insects and honeydew harvested from aphids. Indoors, they get their sugar and protein fix from meats, syrups, jellies, and other sugars.

Leafcutter Ants

Leafcutter ants have sharp jaws that they use to cut pieces of leaf from the plant to bring back to the nests. Once a scout finds a plant it likes it communicates the information to its nest mates—all of its sisters—and they do an intensive strip.

Carpenter ant.

If you have cutters, you won't have to look hard to find their trails. Sometimes they march three or four abreast, going in either direction. One group is going to the leafing area. The ants going back are each carrying a piece of leaf that weighs more than they do.

The leaf pieces are used as the culture medium for a fungus that these ants farm and eat. Several distantly related species of ants are fungus farmers. Each ant farms a different species of fungus, indicating that fungus farming was discovered independently by each species of leafcutter ant.

Fire ant.

Remember that most ant species only devote about 10 percent of the nest's work force to foraging. If food is scarce the colony may devote up to 20 percent of its workers to the job of obtaining food. The other 80 to 90 percent are engaged in activities inside the nest, such as nest building and maintenance, fungus farming, childcare, and military matters. I am inclined to believe that

Argentine ant.

they have some kind of music and comedy as well, because the female workers are sterile. Without a sex life, they must need something to do to in the evenings. Do they have friends among their sister nest mates?

Fire Ants

Red, imported fire ants, native to Brazil, pose a much greater problem in the southern United States than in South America. There are two reasons for this. The first is that the southern United States does not have the same built-in environmental checks and balances that control fire ant populations in their homeland. In South America, they faced predators, parasites, parasitoids, fungal diseases, and other problems that were left behind when they left their homeland.

Local ant predators in the southern United States are species-specific and don't view the imported fire ant as food or a suitable host. The second reason for their success here is that there were just a few fertile females that arrived here. The entire population of imported fire ants is the result of the accidental introduction of just a few queens. Each queen's progeny recognize the entire group as sisters. As a result, nests are actually part of vast cooperative super-colonies that stretch for miles and have thousands of closely related queens. As a result of this cooperation and lack of territoriality among the related nests, numbers in the United States are about seven times as dense as in comparable areas of South America.

Argentine Ants

Argentine ants are native to Argentina, Uruguay, Paraguay, and Brazil. They arrived in the United States on a coffee shipment to New Orleans

LEAFCUTTER ANTS

Leafcutter ant making the cut.

Carrying leaf cuts to the nest.

Tending to food fungus growing on leaves.

in the late 1890s and quickly spread across the southern United States. Like the Conquistadors before them, Argentine ants have rapidly conquered southern and coastal California, sometimes even killing ant species up to 10 times their size. They are imperialists; colonies have populations in the millions with multiple queens and sub-colonies, and different related colonies ally with each other to battle for valuable territory.

Argentine ants are dull gray and about ⅛ inch long. They prefer mild, warm climates and have adapted to urban areas with well-watered gardens. They venture indoors to seek shelter in the rain but often lead the rest of the colony to move in upon discovering abundant supplies of their favorite foods. Argentine ants will eat almost anything; they prefer sweets but consider egg yolks a favored delicacy.

Argentine ants have two stomachs, one for themselves and one to communicate with others. They regurgitate food into each other's mouths, and with the food pass along messages in the form of pheromones. Pheromones are smelled and can be used to communicate anything—attracting mates, work instructions, excitement, and danger. Additionally, pheromones create trails to food sources and distinguish colonies from each other.

ANT CONTROL

There are a number of ways to keep ants out of your garden and off your plants. The first is to prevent them from getting into the garden in the first place, or to cut them off once they start showing up. The second is to sabotage their food supply. Even though there are so many kinds of ants, most species of ants can be controlled through their food system. Understanding that ant species have a food system based on a diet of sugary honeydew is key to eliminating them from the garden. It may be best to approach ants by first trying barrier methods, and if this does not eradicate the problem, proceed to biological controls that affect the ants' food supply.

Creating Barriers

Since ants spend most of their lives as crawling insects, the only way they can get from point A, where they are, to point B, the target plant, is by walking to it. Barriers are a great way to prevent ants from getting to their target. Make sure the garden is well sealed off to ant entryways by using one of these methods. Ant control, especially outdoors, should start with an attempt to convince the ants that your garden isn't that attractive by the use of milder barriers such as cinnamon, or diatomaceous earth. If they can't be persuaded to leave your plants alone, more drastic measures such as boric acid can be used to remove the offending colony.

Food-Based Combat

Barrier methods may stave off an ant colony. However, if it seems that you've only diverted traffic, you may want to change strategies and go for a food-system approach.

Food-system approaches work because they use timing to their advantage. Contact sprays kill ants quickly and may give you a certain satisfaction, but this does little to harm the colony, and is therefore not recommended. Although individuals are short-lived, occasionally living up to 90 days, queens on the other hand are capable of living for years. A fast-acting ant-icide won't do. Instead, use a slow-acting insecticide that the foragers deliver to the nest over a period of time. The poisoned food will be distributed throughout the nest and will eventually be ingested by the queen.

There are a number of baits that use low toxicity chemicals to gradually poison the colony. Almost all ants are attracted to grease and protein, sugars, or both. Since the aphid-herding types are slurping honeydew, we know they're into sweets. Bait designed for sweet-loving ants will lure them.

There are hundreds of brands and formulas of commercial ant baits that mix attractants such as food with the poison. Many popular versions use some form of boric acid as their main ingredient. It is also possible to make your own ant bait or ant brew. The poisons kill slowly so that they are consumed not only by the foragers, but also by nestmates with whom they share the food.

Biological Controls

With the exception of fire ants, there are few biological controls for ants. That is, there are no beneficial insects, parasites, fungi, or bacteria commercially available that can help fight against the ravages of ants. A few commercially available biological controls actually work to successfully eliminate ant colonies. Some organisms, such as the phorid fly, are currently being used by federal and state authorities for the control of fire ants in public lands, but it is not yet available for home garden use. When available it will be a safe solution. Some formulations of parasitic nematodes such as ANTidote are available for home use, but follow-up research shows that the effectiveness of the nematodes varies greatly by ant species and by application.

Indoor Search and Destroy Missions

Pest control books often suggest that you follow a scout ant back to the colony to discover its exact whereabouts. Either these authors don't have a life and take enjoyment in the simple pleasures of tearing apart their homes in search of the hidden ant center, or they are just bluffing. You may occasionally find an indoor colony but it's very difficult, sort of like chasing a black cat outdoors in the dark.

We must use other means of getting to the colony. Remember, only 10 to 20 percent of a colony's population forage for food and all ants except for the queen are infertile. There is a period when reproductive males and females are produced but these leave the colony on nuptial flights to form their own colony. The queen is the ultimate target. Knock her off and it's checkmate in Antsville. Using either barrier or food-based strategies is a better use of your time and energy if your goal is to kill the ants rather than study them.

Ant Solutions Outdoors

Outdoor ants are usually not a problem. However, some species, such as fire ants and harvester ants, can be more than a nuisance in the garden. First, analyze the problem and set specific goals. Realistic goals are to keep the ants off the plants to prevent aphid herding, to keep the ants out of a specific section of the garden, and keep ants from attacking you. It is unrealistic to attempt to get rid of ants in your garden. They are part of the natural ecology, serve a purpose, and are impossible to eliminate.

Keeping ants off the plants can be accomplished by adapting a sticky barrier for outdoor use. Barrier herbs such as bay leaves, cinnamon, and cloves can also be effective indoors and outdoors. Diatomaceous earth can be used to create a barrier but it loses repellency and killing power when wet. It adds boron to the soil. Since plants use only minute amounts of this substance and are sensitive to overdoses, boric acid shouldn't be sprinkled directly on soil.

SUMMARY
Ant Controls

Look in Part 2: Controls for more detailed explanations of the solutions listed below. Part 2 is divided into four sections, each representing different strategies.

Preferred methods are marked with an * asterisk.

Preventive Measures

Keep food waste out of the garden. When developing a plan to deal with an invading ant colony, use their food collection habits against them. Use ant bait with a slow-acting toxin that the ants will take back to the nest.

Barriers & Physical Controls see *Controls: Section 1*

   Baking soda
   Bay leaves
 * Boric acid
   Capsaicin
 * Cinnamon powder
 * Cloves
   Cream of tartar
   Diatomaceous earth
 * Moats
   Potassium bicarbonate
   Repotting
 * Sticky barriers
   Talcum powder
   Vinegar

Pesticides see *Controls: Section 2*

 * Ant bait
   Ant brew
   Baking soda
 * Boric acid
 * Cinnamon oil and tea
   D-limonene
   Garlic
 * Herbal oils
   Hydrogen peroxide
   Insecticidal soap
   Manure tea
   Neem oil
   Pyrethrum

Biological Controls see *Controls: Section 3*

   Flies:
      phorid fly
   Fungi:
 *   *Beauveria bassiana*
   Nematodes:
 *   *Heterorhabditis bacteriophora*

Aphids

Aphids are not considered bright, even for insects. Think of them as the dimwits of the insect world. Their only goals in life are simple: to eat and reproduce. Many people might say, "I know people like that," but there is a difference. Aphids, at some point during their life cycle, reproduce parthenogenetically (no mating required) and give birth to living young, all of which are females.

Aphids puncture leaves and stems to reach the phloem bundle in plants. The phloem is an internal layer of living tissue that works like a circulatory system within the plant. It transports the nutrient-rich sugary sap manufactured during photosynthesis to other parts of the plant where it is used to support cell metabolism and tissue growth. The plant's juices contain a concentration of sugars, but only weak solutions of proteins, which the aphids need in order to grow. To obtain proteins, they suck a lot of juice, refine the protein, and excrete the concentrated sugar solution out of their bodies. This sweet aphid excretion is honeydew.

When plants receive a high-nitrogen fertilization program, they experience enhanced growth. Plants use the nitrogen to produce more amino acids for tissue growth. This means the plants also contain higher levels of protein, which is very attractive to aphids. Aphids that identify higher protein plants enhance their growth rate. High-protein diets also promote egg development in adults, allowing aphids to reproduce more robustly.

Aphids have a hard time living on their own. They are sweet and tender morsels, with no protective armor, no chemical deterrents, and just a few exit strategies. Aphids are sort of like sheep—and they are herded like them as well. Ants tend aphid

At a glance

Symptoms
Disfigured and curled leaves, overall limpness, clusters of bugs on stem and underside of leaves, ant farmers.

Actual size

1/8" • 3 mm

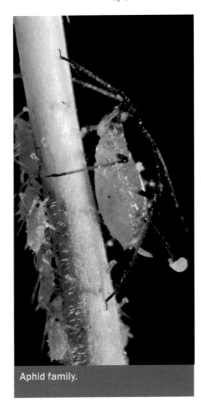

Aphid family.

Aphids have distinctive "cornicles," which look like tailpipes extending from their abdomen.

aphid (*Phorodon humuli*) spends its entire life on cannabis as its sole host. This aphid is present in North America, but is not common.

Many others use a primary host in the winter and migrate to a secondary host during the summer. The hop aphid (*Pardon cumuli*) prefers cannabis as a secondary host after overwintering in plum and peach trees. It can migrate great distances looking for acceptable hosts, and is more selective than the other aphids willing to consider cannabis as an acceptable option. This isn't surprising, given that hops is marijuana's closest relative in the plant world. It is the only other plant that definitively shares the taxonomic family Cannabaceae with cannabis. Because the Pacific Northwest has a long tradition of hops production, and produces the majority of commercial hops in the United States, hop aphids are more common throughout this region. Globally, hop aphids are more likely in hops growing regions, which are optimally situated between latitudes 35 and 55 degrees north or south of the equator, which includes most of the United States.

Three additional species, the green peach aphid (*Myzus persicae*), the black bean aphid (*Aphis fabae*), and the cotton or melon aphid (*Aphis gossypol*) are the most likely to select your crop as a suitable secondary host. However, many other aphid species also find the plants luscious. These species are less discriminating in their selection of plants, and will settle for cannabis if that is what is available.

APHID ANATOMY

Aphids are small, pear-shaped, soft-bodied insects usually less than ⅛ inch or 3 millimeters long. There are thousands of species that vary in color from light green and yellow to black or

flocks, protecting the little suckers from predators and environmental calamities. They move the aphids around to better vegetation and milk them either by squeezing them or stroking them with their antennae to obtain the honeydew they exude. Predators that try to poach the flocks are attacked by the ants, allowing the aphids to suck plants unmolested. There is a good chance that you will find ants if you encounter an aphid infestation.

Some species of aphids are host-specific; they enjoy only one type of plant. Only the bhang

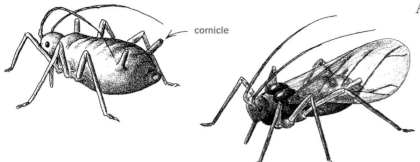
cornicle

Aphids are easy to detect. Turn over undersides of leaves regularly. If they are present, you'll see them. If you feel any honeydew, look for the source and for ants.

Aphid Biology

Aphids are most active between 68 to 85 °F. Their metabolism slows at lower temperatures and they cease their feeding activities to protect themselves from excess heat at high temperatures. They prefer humid, even rainy, weather over dry, warm weather. During favorable weather conditions, aphids are born alive. They are all female and

brown. Some are covered with wax or "wool" made from webbing they secrete, and others have unique distinguishing features.

The feature common to all aphids that distinguishes them from other insects is a pair of "cornicles," which look like tailpipes extending from their abdomen. Almost all pictures of aphids show their backs, legs, and portions of their heads and antennae. Variations in these structures are used to identify the pests. The pictures may not show the cornicles. They also rarely show the stylet, or proboscis, the hypodermic needle-like mouthparts the insect uses to puncture plant tissue and suck its juices. The tube is a composite organ that includes a salivary tube to inject anti-coagulants and a feeding tube.

CHECKING FOR APHIDS

Aphids colonize the stems and undersides of plant leaves. Some species, such as the black bean aphid, are quite noticeable because their color stands out from the plant. Others, such as the green peach aphid, are often colored spring green and blend in with young leaves.

If you have an infestation, you may notice mummified aphids. An egg-laying wasp has parasitized them. This is a good sign that natural enemies are combating the aphids.

Aphid Adaptations

Aphids have been called miniature cows, plant vampires, and insect vegetables, but some aphids show social aspects. There are several families of aphids that induce plants to form galls, or outgrowths in the plant tissues, in which the aphid and her clone daughters live. Aphids of one genus, *Pemphigus*, produce some daughters that are born as a special soldier class. They have stronger legs and more weapon-like stylets than other aphids, which they use to pierce enemies. The soldier class does not mature fully, but stays in an adolescent state. When the gall is attacked by a predator, such as an aphid lion or a wasp, the soldiers mob the attacker, covering its body and using their stylets to puncture it to death.

Soldiers are tested for kinship within a gall. About 40 percent of them turn out to be unrelated. They were less likely to protect the gall than the related ones were. It figures. The unrelated aphids had no genetic interest in the outcome of the fight, so their individual lives were of most interest. The related aphids had an interest in the colony's continuation, even if they should die.

Aphid convention on a stem. They like to cluster in herds.

are smaller versions of the adults. Their growth and maturation is determined by temperature and food. Since they are cold blooded, their rate of metabolism is determined by environmental temperature. Aphids grow slower at 68 °F, or 20 °C, than at 80 °F, or 26 °C.

Aphids have a complex reproduction cycle. They can reproduce asexually or through sexual reproduction. This varies by seasonal cues as well as temperature and food availability.

Juvenile aphids look like small adults when they first hatch. They molt four times before reaching maturity, in eight to 12 days, and start to bear 50 to 100 live young. Mature winged and wingless-stage aphids produce up to 10 live young a day, although a more realistic figure is about five.

Within days, this generation of daughters is bearing more kids. In late spring they start giving birth to winged females who use wind and wing to find another type of plant (the secondary host) to feed upon. The number of young produced

varies by species and the life cycle stage of the mother (hatchling, winged migratory, wingless, or sexual), but at all stages they are incredibly productive. Individuals live for three to four weeks in protected conditions, but in a garden environment, they usually live a shorter life.

Aphids can travel great distances on air currents so they can easily colonize new areas. Once the wind has taken them aloft, they respond to different wavelengths of light. At first, they are attracted to deep blue. Once traveling horizontally with the draft, they seek orange-yellow-green light reflected by plants. They are particularly drawn to bright colors that contrast with the background, especially yellow. They prefer sparse rather than dense vegetation and tall plants rather than plants with a low canopy. Once they get closer to target plants, they sense them chemically. They land and start producing wingless young unless they encounter crowding or unfavorable conditions, in which case they produce more winged daughters.

In the late summer or fall, another winged generation is produced consisting of both males and females. They return to the primary host

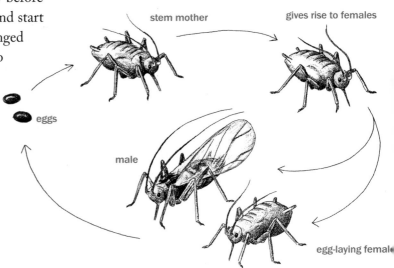

stem mother   gives rise to females

eggs

male

egg-laying female

where they mate. Individuals of some species under some conditions seek shelter in trees or other protected areas. When the weather warms, they are ready to resume eating and reproducing.

Once an aphid has landed in a greenhouse or indoor garden, or if it lives in an area with mild climates, the cycle gets much simpler. There's plenty to eat all the time and no need to worry about the weather, so the insects stop producing migratory winged and sexual stages and just produce generation after generation of the wide-bodied, pear-shaped individuals you may encounter on your plant. Only when the population has reached a density that the local plants cannot support will they start producing winged adults that can fly off to greener pastures.

Disease Transmission

Let's say a winged aphid has inserted its thin stylet into a plant leaf, located the phloem bundle, and taken a drink. Some viral material from the aphid is mixed into the plant's juices. The aphid then moves to another plant and takes a quick suck, decides it doesn't like it, and moves on. If this first plant has an infection, then successive plants the aphid feeds from may become contaminated as well. Its probe into a second plant's phloem layer with the stylet may have delivered a bacterium, fungus, or virus to the plant in several ways.

First, viral material could have stuck to the stylet itself and then jumped ship with the new injection. Second, the plant and the phloem have an interaction much like a mosquito and its prey. When the stylet punctures the phloem, plant cells in the area immediately react by producing a co-agulant to thicken the juices, which are under pressure and flow freely from the wound. This

Why Aphids Suck

1.) Aphids can suck the life out of your plants. Just one aphid can start an infection that has a logarithmic growth rate. Within a few weeks thousands of mothers, aunts, and sisters will be out to eat, and they're all vegans. While a single aphid has little effect on a plant, thousands of little stylets puncturing the plant and sucking its energy-carrying juices weakens and then kills it. The first symptoms are disfigured and curled leaves. This progresses to plant limpness.

2.) Aphids are vectors for hundreds of diseases and can quickly cause an epidemic. They transfer viruses, bacteria, and fungi from plant to plant.

3.) Aphid excrement, honeydew, attracts ants that can be a problem in the garden by harassing you and nesting in root areas, even in containers.

4.) Honeydew is a growth medium for sooty mold. This causes necrosis of leaf parts by hindering photosynthesis.

Leaves that have been killed by aphids.

slows the aphid's feeding through the thin tube. To prevent the activity of the coagulant, the aphid injects its saliva, which contains an anti-coagulant. It may also contain an infectious agent, or in some species it induces the plant to produce a gall.

CONTROLLING APHIDS

Aphids are easy to control. Aphids are difficult to control. It depends on the circumstances. Even though aphids lack protections, have many natural predators, and are vulnerable to some diseases and environmental changes, they reproduce prolifically and bounce back well, leading them to receive the dubious honor of causing the most damage to the world's agricultural harvests. For this reason, many chemical formulations have been invented to kill them. Humans have attempted to outsmart the aphid, which would not seem that hard since aphids aren't that smart to begin with. But these pernicious herbivores have a weapon on their side that may prove more useful than brainpower—genetic adaptability. To date, the results of the battle between humans and aphids have been mixed.

Over the past eons, aphids (and insects in general) have developed tried-and-true ways of overcoming adversity. First, they are willing to sacrifice themselves in great numbers. Their prolific reproduction rates make up for high mortality rates. Second, they use reproduction as their learning kit. Rather than inventing technology, they live it. Several insecticides were very effective when first used. A few years after they were introduced, insects developed resistance to the poisons. Insects have the genetic capacity to deal with unique, extremely potent toxins. In the insect world, chemical warfare is part of life. Plants, fungi, bacteria, and all kinds of other creatures use chemicals either to protect themselves, or to kill prey.

Take plants, for instance. Plants can't run. Instead they use chemical protection to stop predation. They use other chemicals in their arsenal to thwart insects. Glucosinolates are a group of enzymes designed to deal with specific predators. When attacked, the plant produces the appropriate enzyme, which mixes with other enzymes in the plant's cells to create toxins, repellents, and hormone stimulants that interfere with insect maturation and reproduction. Aphids must learn to detect poisons before they ingest them, and to detoxify them if they are ingested. Aphids are guinea pigs in their own ongoing scientific research. Some aphid species have been more

The Great Green Aphid Award— Presenter's Speech

My Fellow Aphids:

Our illustrious leader and breeder, the Great Queen Green Aphid, first predicted an onslaught of chemical warfare by humans more than 1,500 generations ago. Nobody believed her, or believed that we could overcome the novel mental prowess of humans. But the Great Queen was right.

It wasn't easy and it has taken a lot of sacrifice by trillions of individuals who gave their health and lives in order to find cures for the many chemicals humans have thrown at us: organophosphates, pyrethroids and all the hormones, growth regulators, and anti-feeding agents. The first survivor and her immediate successors didn't fare well. They lost their health for science. But thanks to our tremendous reproductive potential, we've had the freedom to experiment with our DNA and find new pathways to salvation. The insect souls of generations past must be gratified to know that they were the progenitors of many new multi-resistant races of aphids.

Our courage to withstand the sprays, systemics, and even the contact-growth regulators, has shown the mammalians who has the wisdom of history. We are stronger for the stress and adversity we have endured, because with it came new knowledge of how to face the challenges of the future. Remember, sisters, we don't have to learn how to do it: It's hardwired into our genes.

successful at adapting than others. The green peach aphid holds the honor of having the most pest resistance of all insects. The cotton/melon aphid is not far behind.

APHID CONTROLS

Aphids are the favorite food of many insects and other small creatures because they lack crunchy armor or unpleasant tasting chemical protectants, are sweet and tender, and since they live in colonies and are slow and sluggish, they are an easy snack food to find.

Aside from predators who want to eat them, aphids suffer from other attackers. Parasitoid wasps use aphids as an incubator for their eggs. They deposit an egg inside the aphid. The resulting wasp nymph uses the aphid as a food source, eating it from inside out, killing it.

Various fungi attack aphids, too. A whole population can die from a virulent fungus in a matter of hours. This usually happens when the weather is damp and a bit on the cool side. The aphid corpses turn reddish or brown and have a shriveled texture with fuzz growing from their bodies.

Wings on a green peach aphid indicate it belongs to a migrating generation.

Aphids being harvested by ants for their honeydew.

Controlling Aphids Indoors

Indoors and greenhouse aphids have an easy life. They don't have threats from weather, and since they have sneaked a backstage pass into a mostly predator-free area, they don't suffer losses to these relentless killers. Without the natural pitfalls they suffer in nature, aphid population growth reaches exponential proportions. Since the balance of nature isn't operative indoors, the gardener must intervene. There are several treatment choices to select from.

Biological Controls

Professional gardeners often use parasites to control aphids when there is an outbreak that hasn't reached epic proportions. Beneficial predator insects are recommended for heavy infestations. However, this may just be a prejudice caused by the subtlety of parasites as compared to the aggressive moves of the predators. The predators spend a portion of their life eating and killing aphids. Close up, their actions can be as vicious and dramatic as an alligator's. The parasites just inject an egg into the aphid, which hatches, and then the young eats the aphid alive until it emerges *Alien*-style from the mummy. Not quite as dramatic, except when the newborn crawls out of the aphid corpse, but every bit as effective.

Some of the predatory biocontrols that are particularly effective against aphids are: lacewings (the larvae are called aphid lions because of their particular taste for aphids), aphid gall midges (*Aphidoletes aphidimyza*), predator wasps (*Aphidius colemani* and *Aphidius matricariae* wasps), lady beetles, big-eyed bugs (*Geocoris pallens* and *Geocoris punctipes*), and minute pirate bugs (*Orius* sp.). The *Beauveria bassiana* fungus also works.

Barriers & Controls

Barriers may improve control over aphids. The most effective types must account for the aphid's small size and ability to fly. Air-intake filters screen aphids from entry into the garden. If they are already in the garden space, the repellant capsaicin may discourage them from sticking around. Aphids can also simply be vacuumed off of plants to reduce their numbers.

Outdoors, row covers eliminate or reduce aphid populations by preventing them from entering.

Pesticides

There are many commercially available pesticides effective against aphids. The best options are insecticidal soap, oils (herbal, horticultural, or neem), and commercially available garlic solutions. Pyrethrum can also be used and may be an especially good choice if the garden requires treatment for multiple pests; however, because pyrethrum is an allergen for some people, it should be used with care. As always, check any commercial products' ingredient list for unnecessary toxins and be sure that it lists its use for food-grade plants.

You can also design your own pesticides from safe household ingredients, such as the citrus oil, D-limonene, or garlic. See the recipe for a garlic spray under Garlic in Controls: Section 2.

Controlling Aphids Outdoors

Aphids do not appreciate extremely hot weather. As the temperature rises above 85 °F or 30 °C, they slow down. They suffer fatalities above 90 or 95 °F (30 or 35 °C). Entire aphid populations are knocked out by extreme heat. One exception to this is the cotton aphid, which can withstand very hot weather.

Aphids do not count marijuana among their favorite food sources, but ants seem to like using marijuana for aphid herding. This means that ridding the plants of aphids often entails getting the ants under control. After the ants are gone, eliminating the aphids is fairly easy.

Sometimes aphids must be controlled outdoors. This can often be accomplished using a gardening strategy that is just about the cheapest and easiest possible: by spraying them off with water. If this simple solution is not enough to bring outdoor aphids under control, consider one of the pesticide solutions listed in the summary, or perhaps adapt a sticky barrier for outdoor use.

APHID CONTROLS

Look in Part 2: Controls for more detailed explanations of the solutions listed below. Part 2 is divided into four sections, each representing different strategies.

Preferred methods are marked with an * asterisk.

Preventive Measures

Monitor and check the plants regularly for aphids—at least twice weekly during the vegetative stage. Once leaves are distorting and curling, control is more difficult because the curled leaves shelter aphids from insecticides or natural enemies. Watch for ant colonies that may be herding aphids onto your plants. Most species of aphids cause the greatest damage when temperatures are warm but not hot (65 to 80 °F).

Barriers & Physical Controls see *Controls: Section 1*

* * Air-intake filters (for indoor gardens)
* Aluminum foil
* Capsaicin
* Heat
* Insect netting
* Sticky cards
* * Vacuuming
* * Water spray

Pesticides see *Controls: Section 2*

* Azadirachtin
* Cinnamon
* * Cinnamon oil and tea
* * D-limonene
* Garlic
* * Herbal oils
* Horticultural oils
* Hydrogen peroxide
* * Insecticidal soap
* Neem oil
* * Pyrethrum
* Sesame Oil

Biological Controls see *Controls: Section 3*

Beetles:
  Lady beetles
Bugs:
* * Big-eyed bugs (*Geocoris pallens* and *Geocoris punctipes*)
* * Minute pirate bugs (*Orius* sp.)
Fungus:
* * *Beauveria bassiana*
Lacewings:
  aphid lions
Midges:
* * Aphid gall midges (*Aphidoletes aphidimyza*)
Mites:
  Iphiseius (Amblyseius) degenerans
  Phytoseiid
Wasps:
* * *Aphidius colemani* wasps
* * *Aphidius matricariae* wasps

Outdoor Strategies

Remember, aphids are a natural part of the outdoors. Be sure they are a pest, and not just a fellow resident of the area. An aphid problem might signal an ant problem. Since ants herd aphids, be sure to address them as well. Adapt the above indoor barrier and pesticide strategies for outdoor use.

Caterpillars

At a glance

Symptoms

Wilted/dead plant material above plant base with signs of bored stems; signs of cut leaves or stems; sudden increases in signs of eaten foliage; caterpillars or larvae in soil at base of plants; signs of silk "tents" similar to spider web visible in plant leaves.

Caterpillar on leaf petiole. Notice damage on upper left.

Caterpillars are the Cinderellas of the pest kingdom. They are an undeniable part of our fairy tales and myths. From early childhood, we hear stories of the caterpillar's metamorphosis. Born to weave its own chrysalis, each caterpillar climbs inside its self-made silken chamber and, like a magician in a disappearing act, closes itself into the cocoon, only to emerge utterly changed from its nondescript form to a winged beauty, a butterfly (or moth, or skipper). Thus, the caterpillar story offers a perfect analogy for transformation in all its possible forms, and encourages us to consider our own ability to transcend our limitations and remake ourselves anew.

Perhaps it is this early identification with the fairy-tale caterpillar's story that leads us humans to feel benevolence toward these creatures. Or maybe it is more a matter of civil interrelations. As an almost entirely (99 percent) vegetarian assemblage of the pest kingdom (the other 1 percent eat insects), caterpillars pose no direct harm to humans—no sting, no bite, no poison for the most part. A few possess venomous spines that break off if brushed against, causing excruciating pain. They are also more chivalrous, failing to annoy us in the way many other pests earn a negative reputation. Caterpillars are downright polite in human terms: no buzzing or noisemaking, and no sudden scuttling over the skin without invitation. They lack a legacy of pestilence or plague that other notorious pests bring to mind at their mention. Finally, let's face it: Caterpillars are cute. They are nearly pet-like in their cuteness, at least in the world of insects and bugs. As such, diplomatic relations between caterpillars and humans are marked by a long-enduring harmony.

Caterpillars, with their inspiring story and polite behavior toward humans, may not even come to mind when thinking of "pests." Even though caterpillars are a favored member of the world of creepy-crawly things, they are still an unwelcome sight in the garden, and can quickly become a problem for a gardener who fails to see their potential as a threat.

Caterpillars are voracious eaters, and may make a feast of plants at many stages of development. There are a large variety of caterpillars, although only a small proportion of them have any interest in a diet that includes cannabis.

GETTING TO KNOW CATERPILLARS

Caterpillars, or more generally, the Lepidoptera order, are a highly diverse lot. The name Lepidoptera is derived from two words: *lepido* meaning scaly, and *ptera*, which refers to wings. They are second only to beetles (order Coleoptera) in their multitude, comprising nearly 180,000 species in 126 families. Five families of Lepidoptera are butterflies, one family becomes skippers, and the other 120 families worldwide become moths.

The Life & Times of a Caterpillar

As we all know, caterpillars don't begin or remain as caterpillars. They transform into butterflies, skippers, or moths. Lepidoptera members undergo a transformation process called holometabolous, which means they undergo a complete change (*holo* means total and *metabolous* means change or metamorphosis). Holometabolous starts with an egg, then proceeds to the caterpillar or larval stage, through the pupal phase or cocoon, to their final adult winged form.

The Lepidoptera order is not the only ho-

lometabolous group. Other well-known holometabolous insect groups are the Coleoptera (beetles), Hymenoptera (ants, bees, wasps, and sawflies), and Diptera (mosquitos, gnats, and midges). The major insect orders have larvae with different common names. For instance, moths, butterflies, and skippers have larvae which are usually called caterpillars. Fly larvae are nearly always called maggots. Beetle larvae are often referred to as grubs.

Even though caterpillars are but one stage in the Lepidoptera life cycle, they spend two weeks to a month in this form; for most, it will be the longest phase of their lives. As caterpillars, their lives are about one thing and one thing only: eating. Caterpillars are the fast-growing children of the insect world, being just a few millimeters in length after hatching. By the time they prepare to pupate, most increase their size by a thousand-fold; some grow by ten-thousandfold from their original size!

Throughout this growth, caterpillars molt, or shed their exterior, much like a snake shedding its skin, four or five times before moving along to the pupa phase. It is upon shedding their exterior that they grow, and these intervals between moltings are called instars. A caterpillar's life is not so much organized around days and weeks as it is counted in terms of instars.

As youngsters, caterpillars are not yet concerned with dating or mating. They have only two goals. The first is to not get eaten. As we will see below, their coloring and a few other defensive tricks help them avoid predators. The second is to grow, and in order to do this, they have to eat … and eat, and eat some more. This is the last hoarding feast they experience. Once they leave the caterpillar phase for the pupal stage and emerge from their cocoons as moths, skippers, or butterflies, their binging days will be behind them. In their new forms, they will proceed with the appetites of supermodels, renouncing all solid foods in favor of a liquid diet of nectar.

Caterpillar Anatomy

Have you ever looked at a caterpillar's face? If you have you will notice that their heads are bisected and come together like two lobes along the center line of the head. Caterpillars have grooves in their "faces" where these halves come together, forming a vertical triangle in the middle tops of their heads. The coloration, depth, and separation where this line comes together between the "eyes" differs by species, and helps differentiate similar types of caterpillars.

All caterpillars have a well-defined head, followed by three thoracic segments, then abdominal segments. On the first segment after the head, many caterpillars have a shield, or a protective plate, which looks like a dark spot.

Caterpillars, like all insects, breathe through tiny openings called spiracles that are located on the sides of the first segment after the head, and along each abdominal segment. These branch into the body cavity as a network of tracheae that supply the body with air.

Some caterpillars are able to detect specific vibration frequencies, but on the whole, caterpillars don't see or hear a whole heck of a lot. They "see" through six tiny simple clustered eyelets, called stemmata or ocelli, which are located on each side of the lower portion of their head. These eyes mainly detect light but probably don't produce high-resolution images in the same sense that we think of seeing. In order to compensate for their rudimentary sight, caterpillars wiggle their heads from side to side to aid in judging the distance of objects, and supplement this with information they gather using their short antennae. The vibrations that some caterpillars hear correspond more to wing flutterings of their predators than anything related to the human world.

Most caterpillars also have setae, or hair-like projections. These sensory organs help the caterpillar feel its way along in its Mr. Magoo world. Even smooth caterpillars have at least six primary setae, while other types are coated with fuzzy secondary setae. Some also have a bumpy quality that is created by short, stout setae. The little hairs or setae aid in the peristaltic motion that helps them stick to surfaces and avoid backsliding. Differences in setae also aid in the identification of different types of caterpillars.

Most caterpillars range in size from ½ inch to 3 inches in length at maturity. While they may lag woefully behind humans in their sensory capacities, caterpillars far outnumber humans in numbers of muscles: They have 4,000! By comparison, humans have between 650 and 850, depending on who you ask. This intense musculature comes in handy for their quirky version of locomotion.

In order to move, caterpillars contract the muscles in their rear segments, pushing "blood" forward into the front segments and causing the torso to elongate, nudging the caterpillar forward. Caterpillars have both true legs and prolegs. The number of each varies, but the true legs are paired on the thoracic segments, while the prolegs are paired on various abdominal segments used for crawling. The prolegs are different because they are not jointed. They also have hook-like structures at their ends that distinguish them from true legs. These hook-like legs are known as crochets, because they look like the crochet hook used in crafts. Crochets are unique to the caterpillar—no other order has them.

Different families of Lepidoptera have different numbers and positions of prolegs, giving slightly different tempos to their perambulations. For instance, the caterpillars known as inchworms or loopers—or more formally as geometrids—have the characteristic cartoonish walk in which their entire body folds in half, then lays flat again. This is because they lack prolegs on all but their hind ends. Other caterpillars that possess more prolegs may appear to have a smoother, continuous, wavelike motion as they move.

Perhaps of all their characteristics, the one that concerns the gardener the most is how and what caterpillars eat. Caterpillars are eating machines. Their tubular bodies mostly consist of intestines that can quickly and efficiently digest plant material. The various families of Lepidoptera have specialized mandibles, or jaws, that are like tools customized to reduce plant materials into meals. Varieties of Lepidoptera may specialize in their diet. They mostly feed on live leaves, fungus, decaying leaves, pollen, spores, or stems. A few feed on plant and animal detritus. Some prefer certain plants to others.

Many caterpillars have nocturnal eating habits. For example, cutworms tend to spend the daytime hours hiding at the base of the plant, and only come out to eat at night. Others are daytime feeders in their earliest instars but as they develop they may shift to eating at night, partly depending on the population density of immediate siblings and cousins.

Eating also indicates to the caterpillar when it is time to build its nest or cocoon and enter the pupa phase. At this point, the caterpillar has gained enough size and nutrients to proceed to seclusion before emerging as an adult butterfly, moth, or skipper.

Caterpillars are voracious eaters, and may make a feast of plants at many stages of development.

Caterpillar looper on cannabis leaf, and its damage.

It is in the adult, winged form that Lepidoptera mates. Butterflies and moths have no interest in foliage. In their transformation from caterpillar, they've traded in their mandibles for a proboscis, a sort of tubular snout designed for slurping nectar. They are no longer a direct threat to plants in this form, but adult females mate and lay eggs on or near the leaves of the preferred caterpillar diet, so that their progeny will wake to a sufficient feast. Butterflies and moths prefer specific host plants, and they can "taste" that they are on the right plant using their feet. They may also hide the eggs underneath the leaves in order to protect them from the elements. Each female lays between a few and a few hundred eggs, depending on the species.

Caterpillar Habits & Defenses

The four-stage process of metamorphosis is one of the successful strategies that allow the Lepidoptera species to thrive. While each stage may have a weak link, the evolution through these phases means that the Lepidoptera species relies on a different habitat during the immature caterpillar stage than in adulthood as butterflies or moths. As adults, the Lepidoptera can fly long distances to reach new potential habitats in order to preserve the species.

Because caterpillars make a protein-rich meal, many animal species are happy to include caterpillars among their snack food. In order to discourage such snacking, caterpillars have developed a number of rather ingenious passive defenses.

The first is coloring. Caterpillars often come in colors that are camouflage for their habitat, whether it is matching the host plant itself, or mimicking items such as bird droppings or branches that might be found in the immediate surroundings. In addition, some caterpillars have developed costume-like coloration, such that they appear to have eyes on their posteriors. If a predator makes a move on them, this gives them a better chance of escape.

Silk is another defense mechanism. Some caterpillars use a line of silk to lower themselves when the branches of their host plant are disturbed. Others create a makeshift tent from leaves. They roll the leaves down, using a bit of silk to secure the sides so they can climb inside to eat in private, away from the eyes of any predators. Some species of Lepidoptera use a similar tactic, spinning silk around leaves to form a shelter for the pupal stage rather than forming a cocoon or chrysalis like their cousins.

Some caterpillars take a more active approach in defending themselves. They regurgitate digestive juices, which are acidic, to spit in the faces of their enemies. Yet others have developed glands that emanate a bad smell to discourage any would-be snackers from choosing them. Some hairy caterpillars have developed setae with detachable tips that lodge in anything that touches them. While most only cause a mild irritation to humans, this can discourage a continued attack. Some caterpillars eat plants that are poisonous to other animals. While the caterpillars are unharmed, the plant toxins in their system make them poisonous to eat. Finally a few caterpillars create protective associations with other insects, most notably with ants, rewarding the ants with food in exchange for protection.

Cannabis-Eating Caterpillars

Common caterpillars found munching on cannabis are also known by other names: Cutworms, cabbage worms, loopers, and stem borers are all species of caterpillars that may try to take up residence in a cannabis garden. While much more likely to be found in an outdoor garden, they may also make their way inside a greenhouse or indoor growing area.

The danger caterpillars pose to a garden is their insatiable appetite for leaves and stems. A young garden can be decimated by caterpillars due to the physical damage they cause to fragile or developing plants. They can completely defoliate plants and leave them weak and susceptible to disease. Some caterpillars eat buds as well as leaves and stems, gnawing through the greenery at an alarming rate that only a pest of their size and single-minded determination could achieve, leaving only caterpillar droppings and fungal infections in their wake. In the world of garden pests, caterpillars are easy to spot but due to their nocturnal habits, gardeners may only see the telltale signs of their presence—munched and chewed leaves, holes in leaves, holes bored into stems, or other signs of rapid snacking. If the invaders are tent caterpillars, then curled leaves entwined with silk may be their calling card. Of course, other pests, such as stem borers, can cause similar looking damage, so the best way to verify caterpillars is by actually seeing them on the plants.

GETTING RID OF CATERPILLARS

One of the best courses of action in a garden is preventing the conditions that can lead to a pest invasion. Keep the garden space neat. Clear out debris, dead leaves, and pruned branches to eliminate items from the caterpillar menu. Keep unhealthy plants away from the main garden. Clean

and disinfect tools and equipment and change into clean clothes and shoes when entering the garden to ensure that you didn't just become the fast lane for pests to enter the garden. Even so, there is no stopping a fecund female from depositing her eggs on your plants. The eggs hatch in six to 10 days.

If you notice the presence of caterpillars in your indoor or outdoor garden, the first step is a fact-finding expedition. Be sure you haven't

Caterpillar stem damage.

Corn borer damage to stem.

Caterpillar damage.

simply gotten a single adventurer who has no palate for marijuana before bringing in the big guns. However, it is important to know that caterpillars are not exempt from pest status, and if you notice their presence—or more importantly, evidence that they have taken up residence and commenced a buffet marathon on your plants—swift action on your part may be the only thing that keeps your garden from being stripped of foliage and weakened. While there are a few different types of caterpillars that may make an appearance in a garden of marijuana, most respond to the same deterrents or are susceptible to the same repellants or pesticide treatments.

Pluck Them Away

Because caterpillars can be visually identified and are larger and slower than many potential garden critters, a reasonable first line of attack on caterpillars is to physically pluck them from the plants. Kill them by dropping them into a bowl or cup of water with a normal amount of dish soap added to it to make it sudsy, or just crush them. If you have hairy caterpillars, or you are just squeamish about handling them, wear rubber gloves. Early morning or late evening are prime caterpillar-hunting times. If the caterpillars are nocturnal, stage a night raid on them with flashlights. Camping headlamps also work well for this purpose because they allow both hands to be free for the job.

In addition to removal, the gardener can go one step further by also identifying, removing, and destroying caterpillar nests and eggs. These are fairly easy to spot. The best way to remove them is by pruning the areas to which they are attached. Some types of caterpillars are borers. These may not be on the surface of the plant, because they bore into large stems. Once inside, they may go up or down, but usually they go up toward the foliage and canopy. You can often find where they have entered the plant because the plant material above where they enter wilts, indicating they are there. You can cut the stems open to remove them or just remove the wilted portion of the plant.

While physical removal of visible caterpillars may slow the progress of the all-you-can-eat buffet on the plants, it is not sufficient to create a caterpillar-free garden. However, it is certainly an environmentally friendly approach, since it involves no introduction of chemicals or pesticides of any kind.

Combine Removal With Repellant

Active removal can be combined with strategies of physical blocking to keep more caterpillars from taking up residence. Some of the more effective barrier methods include screening plants with insect netting to prevent adults from laying eggs on plants; wrapping stems with aluminum, cardboard tubing, flypaper, or insect glue on paper at the base of plants; and using cinnamon, cloves, or both as a spice barrier that is unpleasant to caterpillars.

If the caterpillar invasion is more intense, you can combine physical removal with repellant sprays. Capsaicin sprays, derived from hot peppers, and garlic sprays, or both in combination, are highly effective deterrents to caterpillars.

Pyrethrum works as an effective pesticide against caterpillars. Pyrethrum is a substance extracted from flowers in the chrysanthemum family. It is a very effective natural insecticide that also works for other insects such as mites, aphids, leafhoppers, ants, and fleas. Pyrethroids are synthetic versions of pyrethrum and also effective and low in toxicity.

> Some people are allergic to natural pyrethrum and it is also toxic to fish and reptiles. More information can be found in Part 2: Controls.

Blast Them With Biological Controls

One of the most common and widespread recommendations for a safe, food-grade solution to caterpillars is *Bacillus thuringiensis* var. *kurstaki*, also known as Btk. Btk is an earth-friendly, naturally occurring soil bacterium. In use since the 1930s, this treatment is a favorite of gardeners and farmers because it acts selectively, killing caterpillars but poses no harm to beneficial insects, earthworms, fish, birds, cats or dogs, other mammals, or humans. If caterpillars are an annual problem spray the plants with Btk as a preventative. Caterpillars will ingest it, get sick, stop eating, and die.

Trichogramma wasps parasitize the eggs of many species of Lepidoptera. These beneficial insects solve caterpillar problems without the introduction of any chemicals, natural or otherwise. *Trichogramma* has become one of the most popular biocontrol measures by farmers and gardeners around the world.

Cutworms and some other smaller caterpillars may also be effectively controlled using another beneficial insect: lacewings. Lacewing larvae are a true beneficial because these parasitic eaters will devour not only caterpillar eggs and small caterpillars, but also aphids, mites, thrips, and immature whiteflies.

SUMMARY
Caterpillar Controls

Look in Part 2: Controls for more detailed explanations of the solutions listed below. Part 2 is divided into four sections, each representing different strategies.

Preferred methods are marked with an ✱ asterisk.

Preventive Measures

Planting indoors should eliminate caterpillar problems. For outdoor gardens, keep seedlings indoors as long as possible before transplanting to prevent caterpillars from wiping them out. Throughout the year, clear the garden of weeds and plant debris, especially toward the harvest season.

Barriers & Physical Controls see *Controls: Section 1*

    Aluminum foil

    Capsaicin

    Cinnamon powder

    Cloves

✱ Insect netting

✱ Pest Removal

✱ Sticky barriers:
      flypaper or insect glue

✱ Vacuuming

✱ Water spray

Pesticides see *Controls: Section 2*

✱ Black pepper spray

    Capsaicin spray

✱ Garlic caterpillar spray

    Herbal oils

✱ Pyrethrum

Biological Controls see *Controls: Section 3*

    Bacteria:

✱   *Bacillus thuringiensis,* var. *kurstaki* (Btk)

✱   *Saccharopolyspora spinosa* (Spinosad)

    Bugs:
      mirid bugs

    Fungus:

✱   *Beauveria bassiana*

    Lacewings

    Mites:
      Iphiseius degenerans

    Nematodes:

✱   *S. carpocapsae*

    Wasps:

✱   *Trichogramma* wasps

Outdoor Strategies see *Controls: Section 4*

Animals such as birds may keep the caterpillar population in check.

Damping Off (Stem Rot)

Afew days ago, your seedling flats were healthy, green, and thriving. Today, wide patches of seedlings lay toppled over like trees in a hurricane, their stems at soil level girdled and dark, shriveled with a dried, wire-like appearance. The greenhouse menace, damping off, has paid you a visit.

Damping off is a broad term for seed and seedling decline or death that can be caused by a number of fungus species, all either soil or water borne. These diseases are widely distributed throughout the world and occur in both temperate and tropical climates. However, they are especially the bane of greenhouse growers. Excess moisture is almost always the catalyst for damping off epidemics. The moisture could be a result of overwatering greenhouse flats, or from excess water supplied to outside plantings from rainfall or irrigation. The moisture is not the cause; it is just the ignition key that revs the fungus into gear. Unsterilized soil is the most common source of an infection, as most fungi that cause damping off diseases live and overwinter in soil.

Damping off is primarily considered a seedling disease, but it can affect older plants. It doesn't always kill them, but it may cause lesions on the stems or cause the roots to rot, retarding the growth of the plant and lessening its yield.

Damping off affects seedlings at two different growth stages: pre-emergence and post-emergence. First, damping off infects the ungerminated seed as soon as water begins to soften the seed coat. The mycelium of the fungus invades the seed's tissues, absorbing the seed's nutrients. The seed turns soft and mushy and eventually disintegrates without germinating. In seedling flats, a generalized failure to germinate is often attributed to damping off.

At a glance

Symptoms
Failure of seeds to germinate; seedlings that go weak and bend or break close to the base of the stem. This may also affect older plants. Dried-out leaves; plants wither or go limp and topple; stalk ends of cut plants are dark and shriveled; signs of excess moisture or water in garden.

The stem cells collapse preventing the flow of plant juices.

In a post-emergence infection, damping off attacks the newly germinated seedling at the soil line where the succulent tissue is most vulnerable. The initial damage appears as a slightly darkened, water-soaked spot that enlarges rapidly as the fungus grows. As the invaded tissue deteriorates and softens, the cells begin to collapse, making the stem thinner and giving it the characteristic "wire-like" appearance of a damping off infection. The stems and leaves above become too heavy for the weakened tissue to support and the tiny plant wilts and topples over.

Older seedlings that survived the first on-slaught of damping off may not yet be out of the woods. A persistent, low-level infection can eventually cause cessation of growth, yellowing, and finally, death. In these older seedlings, the infection is evident from the small lesions that form on the stem. These lesions eventually "grow" together and girdle the stem. Another common manifestation of damping off is root rot, where the newly developing roots rot just below the ground. Without these new roots reaching out for additional nutrition, the rapidly growing seedling becomes stunted, wilts, and eventually dies.

THE FUNGUS AMONG US

Several different species of fungi cause the majority of damping off infections. These fungi are distant cousins of the mushrooms that grow on the lawn (and show up on kabobs), the baker's yeast that makes bread rise, and the blue in blue cheese. Fungi do not have chlorophyll, the green substance in plants that allows them to make their own food. Instead they gather food from other sources including other living organisms. They grow best in low light, moderately acid environments, and high humidity.

One common damping off fungi, *Phytophthora*, is not found to infest cannabis at all. In fact, cannabis and hemp have been found to act as biocontrol agents against *Phytophthora infestans*, the organism responsible for the Irish potato famine in the mid-19th century.

Until recently, scientists classified fungi as part of the plant kingdom because they have some similarities to plants. Fungi are immobile, they grow in the soil, produce fruiting bodies, and sometimes bear a great resemblance to plants such as mosses. But, thanks to modern genetic techniques, fungi are now grouped into their own kingdom, fungi, directly adopted from the Latin fungus, meaning mushroom. Fungi were separated out from plants partly because their cell walls are composed of chitin, the same material that makes up insect exoskeletons instead of the cellulose found in plant cell walls.

Fungi reproduce by producing spores. Simply put, spores are to fungi as seeds are to plants. Rapidly growing seedlings and water-soaked seed coats produce chemotropic stimulants, which are chemicals that attract the mycelium or the spores of fungi. A fungal spore that was either just released from its parent or was lying in wait for ideal conditions comes into contact with a seed or a new seedling. Thus stimulated, the spore produces a structure called a germ tube or germination tube that penetrates the seed coat or the epidermis of the seedling either through cracks or by pressuring its way in.

Once inside the plant, the fungus begins to produce pectinolytic enzymes that dissolve the plant's cell walls. As these cells collapse, the fungus grows deeper and deeper into the plant, dissolving tissues and taking up the released nutrients through thread-like structures called mycelia, which act for the fungus much as roots do for plants. Soon, the tiny plant or seed is completely consumed by the fungus and dies or, in the case of seed, turns mushy and deteriorates.

THE CAST OF CHARACTERS

Five different genera of fungi have been documented in cannabis as causing damping off: *Botrytis, Fusarium, Macrophomina, Pythium,* and *Rhizoctonia.*

Botrytis cinerea

Botrytis cinerea is the causal agent of another disease that most cannabis growers know as gray mold, but it also causes damping off in seedlings. This temperate fungus prefers high moisture and cool temperatures, and it can be devastating under the right conditions. *B. cinerea* is also a seed-borne infection. Seeds collected from female plants infested by gray mold have an increased risk of damping off. *B. cinerea* also disperses its spores by wind, not water, the common way of many of the damping off fungi.

Fusariums

Several different species of *Fusarium* can cause damping off, the most common being *Fusarium solani* and *Fusarium oxysporum.*

FUSARIUM SOLANI

F. solani also causes *Fusarium* foot rot and root rot. *F. solani* resides primarily in the tropics and is less common in temperate zones. It is an organism of opportunity and enters root wounds caused by other organisms such as nematodes or broomrape. It overwinters in soil as chlamydo-spores, another type of tough, resting spore that can endure environmental extremes. In southern France, the combination of broomrape and *F. solani* is considered the greatest risk to cannabis gardens. *F. solani* is also a hazard to people who come into contact with infected plants. The organism can cause eye infections from contact and respiratory distress if inhaled.

FUSARIUM OXYSPORUM

This *Fusarium* also causes *Fusarium* wilt. The organism, once inside the plant, plugs the water transport tissues of the plant, produces a toxin, and causes the wilting that is characteristic of a *Fusarium* infection. In the 1970s and 1980s, officials in Washington, D.C., explored the possibility of using this organism for the control of illegal marijuana. A virulent strain was collected from Russia and released into a test field by a U.S. Department of Agriculture scientist. The fungus was successful, killing cannabis plants and overwintering in the soil to kill seedlings the next year.

All species of *Fusarium* prefer warm temperatures and can be devastating when introduced into sterilized soil. As with most of the damping off fungi, they thrive in overly moist soil but cannot exist in saturated soil.

Macrophomina phaseoli

In addition to damping off, *Macrophomina phaseoli* causes charcoal rot, a disease of older cannabis plants. This fungus overwinters in soil as sclerotia, which are hard-surfaced mycelium that can endure harsh temperatures and excessively dry or wet conditions. When conditions become more favorable, the spores germinate and colonize young plants. In seedling flats, this results in damping off disease.

In plants that survive this onslaught, the disease moves into the plant cortex and denies the plant water and nutrients. The plants turn yellow, rapidly wilt, and die. The pith inside an infested cannabis stalk is peppered with tiny black spores. Research has found that excessive drought or high temperatures hasten the colonization of plants. This fungus is a fan of warmer temperatures, the optimum being 98 °F. It is commonly found on maize in the Midwest.

Pythiums

*Pythium*s are considered phycomycetes by taxonomists. This means that their mycelia—the thread-like structures that grow through soil or plant tissue—have no septa, or walls, when viewed under high magnification. In other words, the mycelium does not divide up into individual cells. This primitive characteristic is used by plant pathologists as one way to differentiate between *Pythium* and other fungi. There are thousands of species of *Pythium* scattered throughout the soils and waters of the world. They live on dead plant material and act as parasites on the fibrous roots of plants.

Pythium propagates itself through two types of spores. Zoospores are produced in round, sac-like structures on the mycelia of the fungi. When mature, they burst forth from this structure and begin to swim by wagging tiny, whip-like structures that propel them around in the water or moist soil. They swim for a brief time before they settle and germinate, producing a germ tube. This germ tube is the structure that penetrates either the seed coat or the vulnerable stem tissue of a seedling and begins a new *Pythium*. Oospores are the second type of spore produced. They are considered the overwintering

structures of the fungi. These spores also develop in a sac-like structure. Once mature, they may germinate quickly and infest a plant, or they may develop a thick coating and wait for more favorable conditions.

Pythium ultimum attacks both seedlings and mature plants of cannabis in temperate areas, preferring cooler temperatures between 54 and 68 °F. In adult plants, *P. ultimum* manifests itself as root rot of the root tip. The plant does not die immediately, but eventually the infestation causes the above-ground parts to wilt and die due to lack of water and nutrients.

Rhizoctonia solani

Rhizoctonia solani tends to damage seedlings later than *Pythium*. It can devastate seedlings in a pre-emergence infestation or post-emergence. Of the fungi that cause damping off, *R. solani* is the only one that does not require excess moisture for reproduction. See Chapter 16: Root Rot. *R. solani* can be detected using an ELISA test kit.

CONTROLLING DAMPING OFF

The two most important controls for damping off are also basic rules of good greenhouse maintenance: careful watering and sterile soil. Some of the best advice for avoiding damping off is prevention, which is covered in Preventive Maintenance. Of special importance: the segments about avoiding overwatering, providing good air circulation, providing sterile soil, tidying away dead plant material, avoiding early planting, and of course, not planting seeds that have been infested. Soil amendments that increase the availability of silica for plant development help promote the plant's natural fungicidal qualities.

Plants use available silica to strengthen cell walls, which helps them protect themselves against fungal infections.

There are also several more immediate treatments that can help suppress the formation of damping off. Sulfur is an effective anti-fungal. Sulfur and sulfur-copper mixes are often effective in combating the disease's spread. Because of its sulfur content, chamomile tea helps combat the formation of damping off. Charcoal, cinnamon, and seaweed, which contain anti-fungal enzymes, reduce the chances of damping off infections.

Some fungi and bacteria are effective biological controls for damping off. Many work against multiple fungi and can be applied with a variety of methods, including a powdered seed treatment, a root dip, or a soil drench. While they are great preventatives, they are unlikely to control a serious infestation after it has begun. However, they can prevent its spread to other plants. They are listed under biological controls.

ELISA Kits

Testing kits are now available to test plant tissue for the presence of one of the fungi that cause damping off, *Rhizoctonia solani*, before symptoms are evident. This test, called an ELISA test, contains antibodies which, when in contact with *R. solani*, will turn an indicator color, signaling that the fungus is present.

Damping Off Controls

Look in Part 2: Controls for more detailed explanations of the solutions listed below. Part 2 is divided into four sections, each representing different strategies.

Preferred methods are marked with an ✱ asterisk.

Preventive Measures

The best preventives for damping off focus on moisture control and using sterile soil and planting mixes with good drainage. Applying a fungicide to the seeds before planting minimizes post-emergence damping off. Place seeds no deeper than ¼ inch into soil. Do not transplant seedlings outside until they have several sets of leaves and a robust root. Properly aged compost and compost tea help protect plants from all sorts of fungal infections.

Barriers & Physical Controls see *Controls: Section 1*

✱ Charcoal powder
✱ Cinnamon
   Heat

Fungicides see *Controls: Section 2*

   Chamomile tea
   Compost tea
   Garlic
   Potassium bicarbonate
   Seaweed powder or liquid
✱ Silica
   Sulfur

Biological Controls see *Controls: Section 3*

✱ Bacteria:
   Bacillus subtilis
   Streptomyces griseoviridis
   Streptomyces lydicus
✱ Fungus:
   Mycorrhizae
   Pythium oligandrum
   Trichoderma species,
      including *T. harzianum*

Deer

eer are one of the most watched wild animals in the United States. These sleek and swift mammals evoke powerful associations. There is something about the image of a deer silhouetted against the morning mist that stirs a sense of peace and love for nature. Unless of course, that deer happens to be standing in your yard, making breakfast out of your garden.

Spotting a wild deer out the back door was once a rare treat. In fact, a century ago, it nearly became too rare. As a result of unregulated hunting and deforestation in the United States, the deer population dwindled to critical lows in the 1930s when the entire U.S. deer population dropped to 300,000. Some deer subspecies faced risk of extinction if trends continued unchecked. Hunters and conservationists worked together to formulate restrictive hunting regulations, which allowed deer to revitalize their numbers, but as time passed, it became clear that the restrictions did not account for the fast reproduction rate of deer, especially whitetails. The deer population increased exponentially to create an overpopulation problem. Current estimates suggest there are more than 30 million deer in the United States—that's one deer for every 10 people.

It is now not only common to spot deer, but to find them in places where they don't belong. The sight of a deer emptying the bird feeder into her mouth has become aggravatingly commonplace for some people. In many areas of the country, deer have become genuine nuisances that pose a danger to drivers. Every year in the United States, nearly 1.5 million car accidents are caused by deer, causing an estimated 150 fatalities, more than 10,000 injuries, and over $1 billion in damage.

At a glance

Symptoms
Plants stripped of leaves or knocked down and torn up beyond repair; fence damage that indicates deer interest.

Mule deer.

Deer won't usually target an outdoor cannabis garden, but that doesn't mean they won't destroy it under the right circumstances. Knowing what drives deer behavior may help outdoor gardeners plan and protect their plot more effectively against deer destruction.

UNDERSTANDING DEER & THEIR HABITS

Deer live in climates that vary from tropics to tundra, but most large deer species favor temperate "ecotone" regions. These areas are also called edge zones because they are defined as boundary regions where two different types of ecological environments meet. Deer prefer temperate locations with access to both forests and grasslands. Ecotone regions allow deer access to varied resources, including secluded shelter and a diverse grazing selection. Deer also take water availability into account. The preference for ecotone homes helps to explain why deer are attracted to yards that adjoin forested areas.

Deer tend to be solitary seasonally, especially during the summer. Young males, or bucks, group together in "bachelor pods" until the mating season, while females hang out in family groups after breeding season. Older dominant males are mostly loners, establishing and staying in their own territories.

The Rut

Mating season, or "rutting," is the main event that structures the life of deer. Rutting occurs in late fall, usually November or December. Even though deer sex only lasts for a few minutes, deer start preparing for it months in advance. The rut serves as an annual family reunion. It is the only time that the males and females congregate. It also causes males to display dominance in ways that can damage plants.

White-tailed deer and mule deer are both polygynous. This means that males take more than one female mate, although the level of promiscuity varies by species. Bucks are sexually ready long before the does are ready to mate. While this assures that any female who is ready for mating gets a partner, it also guarantees a high level of sexual frustration among the males, which leads to competition for dominance through displays of aggression.

Bucks establish dominance by sparring. Males begin to form antlers in the spring. Antlers are made of live spongy tissue known as velvet. Through much of the summer, they appear to be covered with a mossy substance, especially among mule deer. Through summer, they get larger, reaching full size by the start of fall. However, as autumn progresses, blood flow to the antlers ceases and the bucks begin to scrape away the fuzzy coating by rubbing their antlers on tree trunks and bushes. Bucks shed their antlers every year after the mating season has passed, around December.

Around the same time that their antlers lose blood flow, bucks basically go through a puberty-like transition, during which their sex drive increases without resolution, causing them to attack bushes and trees as surrogate combatants. They also make scrapes. Scrapes are circular areas of torn up earth made by the largest, most dominant bucks, usually under low hanging branches, which the buck also breaks. This ritual is completed when the buck rubs his forehead scent glands against the tree or bush to indicate that he has claimed the territory. Broken branches almost always point down to the scrape, and bucks

Native deer.

may also urinate and defecate in the scrapes as a clear message to other males: Stay out!

Does are only fertile for 24 hours. As they near estrus, bucks initiate courtship by running them, or basically chasing after them to assess their readiness to mate. This gives a whole new meaning to chasing tail. For all the buck's frustration and enthusiasm, actual copulation only lasts for a minute or two. Afterwards, the buck may stick around for a little while, or he may wander off in search of another receptive female. Once a doe has conceived, she completely loses interest in bucks, and is free to move onto the business of incubating and birthing her fawn.

Does carry their fawns for six or seven months and deliver in late spring. When fawns are born, they have no scent. Their coat is the color of their mother's, but speckled with white patches to emulate the pattern of sunshine through leaves. Mother deer only leave their speckled babies alone for a few hours at a time, hiding them beneath ferns or low hanging plants while they browse for food. A fawn stays with its mother for a year, or longer if female.

Communication

Deer are social animals and they have an intricate and fascinating system of communication. Most well known is the white-tailed deer's "flashing" of the tail when danger is sensed. Male and female white-tailed deer raise their plume tails straight up as they bound away when disturbed. Although the classic depiction of both sexes is a deer bounding thorough the woods, tail high, bucks often run with their tail down as often as up. Does, on the other hand, almost always run with their tails high, flashing the white underneath regardless of the situation. Animal behaviorists studying deer have offered the theory that the white-tailed deer's flashing is also a signal to a predator that his presence has been discovered.

Snorting is another signal that all is not well. Anyone walking through the fall woods who has heard the sudden, explosive snort of a surprised deer can well understand how a predator might feel upon hearing it. In order to snort, deer force air through their nasal cavities with mouths closed. To bump the warning status up a notch, they also have a higher-pitched, whistling snort that is made with the mouth open and accompanies a vocalization. Upon hearing this, every deer within earshot will scatter, knowing that danger is imminent.

Deer seem to have a sixth sense about danger, which adds to their aura of mystique. They often know when danger is present even if they are upwind of a threat or out of line of sight from the danger. However, their extremely sensitive hearing and excellent eyesight give deer specialized perception of detecting danger from long distances. Also, when deer sense danger, they stamp a forefoot, often repeatedly, while they scan the woods for the source of their intuition. These tiny tremors travel across the ground and all deer in the immediate vicinity feel the vibrations and are instantly alert. Deer forelegs also contain a pheromonal scent that, when released, inhibits feeding and alerts other deer to acute danger.

Deer have a visual range of about 2,000 feet. Their vision is dichromatic, so they only see colors in the blue and yellow hues, but are red-green color blind. This may help to explain the hunter's choices of camouflage. Humans, by contrast, have trichromatic vision.

Foods & Feeding

Deer are herbivores that eat a wide variety of plants. Their diet is dictated by season. Deer can often be seen mingling through yards, fields, or forests as if they're in a grocery store that's offering free samples. Deer are open-minded nibblers and will taste nearly anything. They graze, eating grains, grasses, or other low vegetation in open fields. They also browse, which means they eat the hardier, woodier plant parts and the raised leaves of trees, plants, and shrubs. These foods—pinecones, tree branches, and elevated plant leaves—are also known as browse.

Given the deer's preferred diet, good locations for feeding are fields and meadows, which are vulnerable locations in terms of exposure. This makes feeding a dangerous task. One way deer avoid danger is to forage at night or in the early morning hours, although they may feed up to three times a day.

Another protective measure comes courtesy of their ruminant stomach. Like cows, giraffes, goats, and sheep, deer have a ruminant four-chambered stomach, which allows them to stuff now and eat later, often known as chewing cud. When deer find plentiful food, they stuff their first stomach with as much food as they can as quickly as they can, and then get back to safety. A grown deer can fill her paunch by grazing for an hour or two in one spot. After relocating to a more protected area, the deer lies down and brings up lemon-sized balls of pre-digested plant material, or cud, to chew at her leisure.

Cud chewing, or ruminating, works because rumen microorganisms in the deer's digestive system break down the food. This adjusts by season. Deer are attracted to the succulent sprouts of grasses and plants in spring. As summer nears, they begin to feed on forbs, or non-grass plants like broad-leaf weeds and clovers. Forbs provide the essential nutrients that does need for birth and milk production and bucks need for antler development. At the end of summer deer begin to browse low-hanging tree leaves and bushes, mushrooms, and ripe berries. Their stomach chemistry allows them to eat some plants that are poisonous to humans, such as mushrooms and sumac. As fall descends, apples become ripe, and a little later, acorns begin to drop. The abundance of an acorn crop can determine the survival or death of a deer community. Beyond ability to survive, the available food supply affects their overall physical condition and reproduction.

Farm and garden crops are special treats and deer can be amazingly aggressive to reach these goodies. They like alfalfa, winter wheat, pea and bean sprouts, and ears of corn, as well as tasty sweet fruits such as strawberries, peaches, and apples. They love roses, impatiens (known among gardeners as "deer candy"), and even the young shoots of rhododendron bushes. They also like white pine and white cedar. They'll eat young cannabis, but it doesn't rate as highly as many of these other cultivated crops.

Deer prefer cannabis in the same growth stage they prefer most plants—during the early vegetative stage. They tend to avoid flowering stages unless other food sources are scarce. Cannabis loses its appeal as more resin develops. This may indicate that deer don't like getting high, although some authors claim that buds upset their digestive system.

These downsides may not keep deer from ripping tops or leaves off plants to determine their desirability. Food pressures may cause them to eat flowers out of desperation. If they feed in the marijuana garden, they will strip plants of leaves, and eat or tear up whole plants beyond repair. Deer may not eat all of your buds in a sitting, but that is little consolation if they trample the rest into the mud before they leave.

What deer eat and what they'll do to get it

In North America, all deer belong to the genus *Odocoileus*, a name bestowed on these mammals by French-American naturalist Constantine Samuel Rafinesque. Gifted but slightly eccentric, Rafinesque was obsessed with discovering and naming new organisms—a quirk that earned him the reputation of glory hound and sloppy scientist. The famous ornithologist John James Audubon, a known prankster, once tricked Rafinesque into publishing the descriptions of nearly a dozen fictitious creatures.

Rafinesque reportedly discovered a fossilized deer tooth in a Virginia cave in 1832. He ascertained that the tooth belonged to a long-dead relative of the whitetailed deer browsing nearby in Virginia's woods and fields and quickly named the genus and species *Odocoileus virginianus*. Unfortunately, Greek was not Rafinesque's strong suit. Intending to name the genus *Odontocoelus*, meaning "hollow tooth" or "concave tooth" he instead misspelled it as *Odocoileus*. Because of taxonomic rules, scientists tend to honor the original names of organisms whether they're correct or not, and so the genus *Odocoileus* was accidentally born.

depends on food pressures such as deer population density, season of the year, and availability of other food sources. This is very important in choosing deer prevention methods. When food is plentiful, deer are not willing to spend effort on hard-to-get digestibles. When food is scarce they are willing to do what it takes to eat the goodies.

Deer damage is recognizable by the ragged tearing and ripping of their grazing. Since they lack upper incisors, they cannot bite vegetation off cleanly like rabbits. In gardens where preventive measures are already in place, deer droppings may also be a reminder to tighten up border security. Deer droppings are small oval pellets about ½- to ¾-inch long that are laid in a cluster. They look similar to rabbit droppings, but deer pellets are always round. Deer poop indiscriminately and often, so they leave behind a telltale trail of their visitation, but this often comes too late, simply confirming who chowed down on the plants.

WHAT'S IN A NAME? NORTH AMERICAN DEER SPECIES

Deer are indigenous to all continents except Antarctica and Australia. The *Odocoileus* genus consists of only two species, which are also the only deer indigenous to North America—the whitetailed deer (*Odocoileus virginianus*), and the mule deer (*Odocoileus hemionus*). Blacktailed deer (*Odocoileus hemionus columbianus*) are categorized as a subspecies of mule deer, but most naturalists believe the black-tail will soon be considered its own independent species.

The white-tailed deer is one of the oldest deer species. The contemporary whitetail is nearly identical to its 3-million-year-old ancestors. This breed resides throughout the temperate zones of North America, from the tip of Mexico up through the southern provinces of Canada. By contrast, the mule deer species is believed to be one of the youngest deer breeds, dating back only 10,000 years. Mule deer are only found west of the Missouri River in the northern part of North America, and west of Texas in the southern region. They prefer the Rocky Mountain region and can be found along its entire span. Black-tails only reside in the Pacific Northwest, from Oregon through British Columbia, and in the southeastern part of Alaska.

Whitetailed Deer

Whitetailed deer are known as the type species for the genus *Odocoileus*, meaning that all other contenders for inclusion in the group are compared to it. These deer have managed to develop opportunistic habits and largely sidestepped evolution.

Whitetailed deer

Whitetails are considered the most nervous and shy of all deer, and their tail-flagging method of communication has earned them their common name. This species accounts for 15 million deer in the United States, about half of the estimated population. These deer are swift and agile: Adults can sprint in excess of 30 miles per hour, leap more than 9 feet high, and bound as far as 30 feet in a single effort. They have the potential to live for up to 20 years, but in the wild, most have a life span of only two or three years, and few live for more than a decade. They reproduce at younger ages than mule deer and have a nearly equal ratio of bucks to does. As an older species, whitetails have a higher resistance to parasites and disease than mule deer.

Because of their highly adaptive vegetarian diet, whitetailed deer maintain a small range and do not migrate seasonally. Does and bucks typically stay within less than a square mile for several years, although bucks tend to have a slightly larger range. They move in response to changes in weather conditions or changes in food availability. They prefer areas with moist streams or bottomlands, and they are amazingly good swimmers, a skill they sometimes use to escape predators. There are some whitetail/mule hybrids, and these deer favor whitetail characteristics. Once they've taken up residence, they tend to establish grazing and browsing trails, and are fairly routine about sticking to established pathways.

Mule Deer

Odocoileus hemionus or "muleys," as they are affectionately known, keep to the American west. These deer do not have white tails, nor do they fling their tails up when startled. Their distinctive feature is the floppy ears, the basis for their

Blacktailed deer

scientific name *hemionus,* which means "mule" in Greek. Mule deer are larger but not necessarily heavier than whitetails. They are known for their bounding gait when on the run because they seem to bounce, springing and landing with all four feet at once, unlike the whitetailed deer, which springs from its back legs and lands front legs first. These deer were not endowed with good swimming skills like their cousins, but they are fast. Mule deer have been clocked at 36 miles per hour in a sprint and they can bound 20 feet in a single leap. While the daily meanderings of the mule deer stay within a small range, this species migrates seasonally, often returning to the same summer and winter locations over several years.

Blacktailed Deer

Blacktails, *Odocoileus hemionus columbianus,* are a subspecies of the mule deer and the smallest of the three types. There are two variants: the Columbia blacktail and the Sitka blacktail. Their name obviously comes from their full dark tail, which is black from base to tip. Their range is restricted to the Pacific Northwest: Oregon, Washington, British Columbia, and southeast Alaska, where they prefer old growth forests because these regions provide shelter and foraging, and resist deep snow buildup in winter. Columbia blacktail coats are darker than their cousins, while Sitka blacktails are the smaller blacktails, yet resemble whitetailed deer in coloration. It is possible, but rare, to see mixed characteristics, since there is minimal hybridization between whitetailed and mule deer species.

Size Matters

Deer are dimorphic, which means that males are about 20 percent larger than females. Like most warm-blooded animals, each deer species tends to get larger as they range farther from the equator. For instance, whitetails in the Florida Keys (*Odocoileus virginianus clavium*) are about one-quarter the weight of the largest whitetails, the woodland and Dakota subspecies (*Odocoileus virginianus borealis* and *Odocoileus virginianus dacotensis*) found at the northernmost part of the whitetail range.

These size differences make sound evolutionary sense. The larger the body mass, the more heat is retained. Smaller body mass dissipates its stores of heat faster. This is important when devising methods to keep hungry deer out of a garden. Keeping a hungry 200-pound deer out requires a different strategy than keeping out a 90-pound deer.

In addition to body size, natural selection governs the size of extremities, such as ears, tails, and legs. These vary by species, but they also vary within species such that the ears, tails, and legs are smaller or shorter in cooler ranges. Deer coloration within species tends to be darker in warmer climes. This is governed in part by humidity. Higher humidity levels lead to darker coloration, whereas hot dry climates, including high-desert mountain ranges, result in lighter-colored animals.

CONTROLS

Deer are persistent. They regularly test their ability to access desirable food sources, and enthusiastically return to places where they found good eats. As a pest, deer size means that even one visit can devastate a garden.

Deer can be real pests for outdoor gardeners. They can jump, squeeze through ridiculously small openings, and are often quite agile about getting in and out. If deer frequent your area, it is important not to underestimate their interest or sneakiness when planning your outdoor garden. If you have an existing garden in place but deer have shown up, you may need to create some barriers or distractions.

USDA studies on deer behavior find that deer develop tolerance to deterrents based on bad tastes, bad smells, strobe lights, and noise/siren systems. Deer also display a circus acrobat's agility and a Houdini-like ability to get in and out of locations. They can jump at least 8 feet, and may reach 12-foot heights when pumped up on adrenaline or otherwise highly motivated. They can squeeze through impossibly small openings, and have been known to sneak under fences that only had an 8-inch clearance. In addition, they are relentless in their willingness to test barriers for weakness and they can learn how to remove netting or protective wrapping on plants.

It is important to understand deer when planning for outdoor garden security. Working with deer tendencies can help. Deer base their determination to get to your garden on a simple cost-benefit ratio of how desirable the plants are versus how difficult they are to access. These animals tend to stick to their regular pathways when heading to reliable food sources, and they are less inclined to jump fences when the enclosure stands in the middle of an open area. Plants become more tantalizing when food pressures mount and the favorite foods of deer are in short supply. This can be a hard factor for gardeners to assess, but it is better to be safe than sorry.

Below are some suggestions for keeping deer out of your crop.

Location

Deer are habitual animals. They follow customary paths when they forage. They also tend to establish a fairly confined range. Steer clear of known deer foraging grounds, and you may sidestep the deer issue altogether by staying out of range.

Repellants

Repellants are most effective where food pressure is low to moderate. These products are often composed of putrid-sounding ingredients, and they do work with moderate effectiveness. Deer are quick learners and once they surmise that a repellent is nonfatal, they will often risk punishment to get to the treat. No repellant works indefinitely, so rotating repellents achieves the best control. Repellents vary in price and labor, and are usually best suited to a small area.

Nobody likes the smell of rotten eggs and deer are no exception. In a study by the U.S. Forest Service, deer repellent products that contained rotten, or putrescent, eggs did far better as a repellent than other products that contained just capsaicin, D-limonene, or other stinky compounds. Commercial formulas include Liquid Fence, Deer-Away Big Game Repellent, and Deer Off.

Name of Repellent	Description
Putrescent eggs	Putrescent egg solids are mild and inoffensive to humans but smell terrible to deer. Commercial products are available, or make your own by blending two raw eggs and one cup of water until pureed. Dilute to 2 liters and spray directly on plants. Reapply after a rain.
Predator odor: cougar, coyote, or human urine, or dried blood products	Can be bought from garden suppliers, hardware stores, and some hunting suppliers. Available in pellets and in liquid.
Milorganite	An organic fertilizer made from processed sewage sludge that, according to research, has some properties as a deer repellent. The manufacturer does not label this product as a repellent.
Capsaicin	Commercial products available, or mash hot peppers and mix with vinegar and water. Apply to the ground around the area. Dilute with more water to spray directly on plants.
Animal waste	Collect carnivorous animal waste (dog, cat, etc.). Place in mesh bags and hang around the perimeter of the growing area.
Deodorant soap	Cut bars of any brand of deodorant soap into strips, place in jars and distribute at deer's chest height around the perimeter of desired area.
Human hair	Collect hair from salons or barber shops. Place in mesh bags and hang at a deer's chest-high level.

The commercial product Plantskydd is also known to repel deer. The Swedish product uses bloodmeal to inhibit deer and was first developed to repel deer from commercial forest plantations.

Dogs keep deer at bay as well. Deer stay away from dogs. A dog in the garden is all the deer need to remind them that they are deerona non grata.

WATER SPRAYS

A number of motion-activated water sprays have recently entered the gardening market as a deer deterrent. These consist of a motion- or heat-sensing device that sprays a 35-foot jet of water in the air when an animal's presence is detected. Harmless to humans and pets, this novel product banks its effectiveness on the randomness of the spray and the ability of the deer to be frightened.

Physical Exclusion

Exclosures are barriers meant to keep something out rather than hold something in. These may be fencing around the entire garden, or other forms of covering individual plants.

ELECTRIC FENCES

In areas where the food pressure is low, a single-strand electric fence 30 inches off the ground might be all that is needed to keep deer out. Deer sense the current running through the wires and know to avoid it. If they are unfamiliar with electric fencing, they learn quickly that the fence delivers an unpleasant jolt of electricity. It is not harmful, but it will scare them.

In areas where the deer population is denser or the deer have learned how to get through a single-strand fence, adding another wire at 15 inches above the ground might help, since deer generally prefer to go under a fence rather than over it.

Another successful electric fence configuration is a two-dimensional design. A single wire is placed at a height of 50 inches. Two additional wires are placed 38 inches outside the first wire at heights of 15 and 43 inches. This creates a two dimensional obstacle that confuses deer. Using metal posts and polywire create a structure that is easy to take down and move.

The most effective electric fence has a high voltage, a low impedance charger, and a good ground rod. Higher voltage chargers require less maintenance because they burn and sever any weeds that grow up to touch them. Chargers are available that use alternative power sources, including solar, AC current, and battery.

Some important points to remember for electric fencing to work are:

Electrify the fence immediately so deer don't have success bypassing it. Once they learn this behavior, it is difficult to re-train them, even after the current is flowing. If they know there is food across the fence they are sometimes willing to accept the jolt.

Fence breaching is very serious because as soon as one deer does it, the information gets passed around and the moveable party is in your garden. Lure the deer to touch the wire by baiting it with peanut butter on aluminum foil wrapped around the fence wire. This may stop the deer from attempting to breach the barrier.

Check the fence regularly for breaks caused by weeds and debris. Deer test the fence periodically to see if it's still operable. Use a fence tester to locate outages.

PERMANENT FENCES

Permanent deer fences are the only sure way to keep deer out of a garden. They will only work if they are high enough to keep deer from leaping

Fencing is the most secure protection from deer predation.

less, shock when deer are baited to touch one of several upright poles positioned around the area needing protection. Invented by a veterinarian to eliminate the need for electric strand wire, which can entangle animals, this method depends on the deer being repelled by the startling jolt and deciding the treat beyond is not worth the pain. Deer become attenuated to it and learn to avoid the baited treats.

NETTING

Depending on the deer's determination, a minimalist method for deterring deer involves placing netting on the plants. If the deer are experiencing food pressure, this barrier may not be helpful as deer are known to remove this type of barrier in times of need. However, if deer are simply walking through the garden on their way to another location, this simple approach may work.

over the top, and they leave no gaps where deer can sneak underneath them. An 8-foot-high metal chain-link fence all around the garden is the best barrier. In some areas plastic fencing can be substituted for metal. Either way the posts holding the fence must be sturdy and securely positioned. Electrifying the top adds to its effectiveness.

In areas with high food pressure, deer will test the fence more. If the fence is too high to breach they may try to use their body mass to topple the barrier. The weight of the deer plays a part in this. In the northern United States, where the deer grow larger, it is more of a concern than in the south, where the deer weigh significantly less.

The worst thing that can happen to your garden is to have a deer jump over the fence to get in but not be able to easily jump out. This results in a trampled crop. For this reason make sure that the fence height is higher than the most athletic deer can jump.

In the last few years, a wireless deer fence system was patented. It delivers a sudden, but harm-

---

SUMMARY
DEER CONTROLS

Preventive Measures

Since cannabis is not a favorite food of deer, even if deer are present in the area they may not be a problem.

Solutions are specific to deer behavior and how they interact with humans. Preferred methods are marked with an ✱ asterisk:
✱ Animals as pest control:
    dogs. See Controls: Section 4.
✱ Fences
   Repellants:
    smells of putrescent eggs and predators
   Insect netting
   Water sprays

Fungus Gnats

At a glance

Symptoms
Swarms or clusters of small gnat insects; yellowing leaves; overly humid garden or damp soil; damp climates; plants with fungus problems such as *Pythium* or *Fusarium* may also have fungus gnats.

Actual size

1/8" • 3 mm

Mycetophilidae: note the humpback formation.

Fungus gnats are often considered harmless because their damage is not apparent. They don't bite, and pose no direct hazard to humans. Their diet consists of fungus and decaying organic matter as well as plant roots. However, complacency about fungus gnats is an erroneous attitude. You should not assume they are benign tenants. They are often an indication that the growing media is too wet—a condition that can lead to a host of fungal problems. Fungus gnats are a danger for a few reasons. When the larvae suck on roots they cause damage that weakens roots, and leaves them vulnerable to infection. Also the larvae are disease vectors.

Fungus gnats look like small mosquitoes that swarm in clusters around the bases of plants and in their foliage. The fungus gnats that like cannabis come from two families of flies: Sciaridae and Mycetophilidae. The most numerous member of the Sciaridae family is the *Bradysia* genus. This genus includes 65 species and can be found throughout the world, but is most common to Europe and North America. In the United States, *Bradysia* species reside in all states, with larger populations in the north and west. Adults in the Mycetophilidae family are humpbacks, although with their small size humans may not notice. It may look more like a rounded head than a hunchback.

Fungus gnats are usually associated with an overly moist environment, since adult fungus gnats rely on a banquet of decomposing plant matter and fungi, both of which are more prevalent when moisture levels are high. They require wet, rich organic soil or other planting media as a home for their eggs. These insects determine where to lay eggs based on the moisture content of the planting medium. They also read chemicals that are emitted by organisms in the growing media. Fungus gnats have preferences when it comes to media type, but they will lay eggs in all kinds of planting mixes including soil and soil-less mixes, rockwool, and the commercial brand Oasis. Less hospitable media such as perlite and LECA, a hydroponic growing media, offer some protection against fungus gnats as they are not suitable for egg laying.

Sterile potting media and rockwool are more susceptible to infestation than soil. Within soil or soilless mixes, fungus gnats are partial to peat moss and composted hardwood bark, and gravitate to these materials. They prefer newly potted media or recently enriched soil, since it contains high levels of fresh compost.

Adult fungus gnats eat garden garbage, not plants. The larvae live in the planting mix and feed on plant roots close to the soil surface. In sufficient numbers, these larvae interfere with the transport of nutrients to the canopy, causing chlorosis or yellowing of the leaves due to lack of chlorophyll. Without chlorophyll plants can't photosynthesize so they cannot produce sugar used for metabolism and growth. Fungus gnat larvae damage the root when they suck on the roots and stems, making plants more susceptible to fungal and bacterial pathogens including *Fusarium* and *Pythium*.

If there's always a little puddle of water in the greenhouse and the soil or growing media is always wet, fungus gnats are more likely to be a problem. This is more likely to occur in the summer for some, as the conditions of fall and winter, with cooler temperatures and fading light intensity, lead to longer intervals between watering and less optimal conditions for fungus gnat development. Others report the summer heat dries out excess moisture in the summer, and fall or spring conditions are more ideal for the fungus gnat.

IDENTIFYING FUNGUS GNATS

Adults of both the Sciaridae and Mycetophilidae families are about ⅛-inch to 1/10-inch long, and range in color from gray to black. Their bodies are slender with mosquito-like legs, beaded antennae, and one pair of clear, veined wings. Fungus gnat larvae are slender maggots, up to ¼ inch in length, with a translucent, whitish body and a shiny black head. It is possible to see food in the maggot's gut through its nearly clear skin. Pupae are found in the soil wrapped in silken cocoons. Adult males are very similar to females. The only difference is that males have claspers on the end of their abdomens, which are used during mating.

Bradysia dark-winged fungus gnat larvae and adult.

Mycetophilidae: fungus gnat larvae and adult.

FROM EGG TO GNAT

The fungus gnat life cycle is a complete metamorphosis, meaning that the insect goes through four different stages of development: egg, maggot, pupa, and adult. Female fungus gnats lay eggs in planting media that is moist and contains organic matter or growing roots. Moist or molding dead leaves and plant debris left on greenhouse tables and floors also provide a good habitat. Females lay between 200 and 1,000 eggs in their seven to 10 days of life. In one experiment, females laid between 21 and 217 eggs in 48 hours.

Fungus gnat eggs are whitish or clear, and hatch into tiny maggots after four to six days. The timing largely depends on the temperature. Indoors, fungus gnats have it made. There are few predators, the temperature is near tropical, and the controlled environment makes it unlikely that eggs will suffer insults such as dehydration. Outside, with the disturbances and accidents of nature, fungus gnats lead a precarious life. The eggs lie right under the surface level before the larvae emerge. Small predators might find these morsels a convenient snack. Some eggs dehydrate, become diseased, or suffer other traumas and stresses that result in death.

The survivors emerge and the just-hatched maggots quickly burrow several inches below the soil surface. Sometimes they follow roots to get to tender young growth, which are easier for them to feed on. They can also survive and mature living on algae, which often is found growing on moist greenhouse surfaces and on top of planting mediums.

The maggots are cloudy, transparent to white, except for the head, which is black. They are legless but use pseudopods, legs made from extended bumps of flesh, to move. Fungus gnat larvae are very hard to detect in the soil but sometimes they can be seen in the bottom of trays when water drains from contaminated containers. Without magnification, they look like tiny caterpillars wiggling in the water.

Outdoors, fungus gnat maggots face formidable obstacles getting to the next stage. Predators are lurking, ready to ambush the maggot and eat, parasitize, or invade it. In addition, the vagaries of nature take their toll. Have we forgotten about diseases? These creatures live in a hotbed of bacterial action and many of them succumb to disease. As a result, very few contestants make it to the next stage, the pupa.

The pupal stage is usually associated with the caterpillar's cocoon. While inactive on the outside, a transformation takes place inside. The maggot spins a soft silk-like skin around its body and recedes from the rest of the world. The tiny

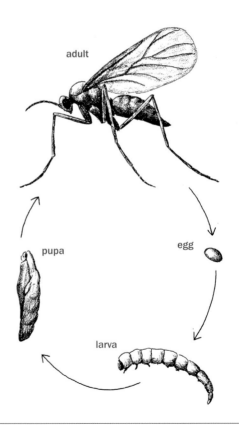

(smaller than a rice grain) tan-to-brown cocoons can be found in the soil, but they are very tiny and not usually noticed. Despite the insect's preference for moist, dark locations, larvae and pupae can and do survive periods of drought. In five to seven days the maggot metamorphosizes into an adult.

When fungus gnats emerge from their cocoons, they buzz around the growing area in thick abundance. They begin courting and consummating their intimate rendezvous. This is the final phase in the gnat's life: adulthood and mating.

One of the nice things about being a fungus gnat adult is that your only responsibility is to engage in sex, and if you're female, to lay eggs constantly for the week that you are alive. Champion egg-layers may lay 1,000 eggs during the week or so that they live. Let's say there's a really healthy, long-lived female gnat that survives 10 days. To get the fertility gold medal, she lays 100 eggs a day. That's a little more than four per hour.

Now think about the gnat couple. After mating, the male is off looking for other females to flirt with. Meanwhile, the impregnated female gives up sex and starts pumping out eggs. To carry on the population, each gnat couple needs only two kids to survive. She's laying 500 to 1,000 eggs. The expectation, nature says, is that only two will survive. It's a perilous journey.

A fungus gnat that dies of old age lived an entire life span of 25 to 45 days. Generations can overlap in greenhouses where the conditions are nearly perfect for reproduction. That's why the population can build up so quickly. The population numbers still follow a natural cycle of sorts. The first two generations will be large, followed by a leveling off and eventual decline as food becomes scarcer.

REASONS TO BE CONCERNED

There are several reasons why you should be concerned about the presence of fungus gnats in your garden. The maggots eat the young root tips and suck the plant juices, which zap plant vitality and vigor while the maggots multiply. That's why with fungus gnats, just as with all other pests indoors, there should be a zero-tolerance policy. One gnat is one gnat too many.

By wounding the roots, fungus gnats leave plants susceptible to attack by pathogenic fungi bacteria and other soil-dwelling organisms. Also, environments that fungus flies can become established in have suitable conditions for other pathogenic visitors such as molds and rots.

Fungus gnats are vectors for plant diseases since they transfer pathogens when they suck on an infected root and then pierce another. They regularly transmit *Pythium*, *Fusarium*, black root rot, and viruses.

These three reasons make fungus gnats an intolerable threat to your garden.

PHYSICAL CONTROLS
Monitoring

Placing yellow sticky traps near the plant canopy attracts fungus gnats. These traps are used to indicate gnats' presence and as a crude measure of population size. Placing cards beneath benches and near air-return areas provides additional information on emerging populations. For more information, see Sticky Cards in Controls: Section 1.

Keep the Garden Tidy

Follow the basic preventive maintenance tips covered in Chapter 1. Fungus gnats are more likely to find an untidy or damp garden attractive. Over-watering is the most common factor that promotes fungus gnat infestations. A simple control is to allow the top of the soil to dry before watering. This is perhaps the best practice. Just don't allow it to dry much beyond the top ½ inch of soil and certainly not so much that the plants wilt.

Eliminate any leaks in the watering system and make sure drainage is complete. A dripping hose or a puddle of standing water often results in an algae-slick area that supports fungus gnats. Remove dropped leaves and other plant debris; decaying plant material can harbor fungus gnat maggots. Inspect and quarantine new plants brought into the greenhouse for fungus gnats and other infections before introducing them into the garden.

Fungus gnat maggots also transmit diseases, so eliminating them from the soil fulfills more than one purpose. Don't reuse soil containing fungus gnat maggots. However, uninfected used soil and planting mixes contain beneficial organisms that live in a symbiotic relationship

with roots. When they are used again the microorganisms are there to greet the new roots. When buying planting mix use an inert formula or pasteurized or sterilized mix. Make sure the bags have no rips or tears

Barriers

Placing a barrier between the plant canopy and the planting medium prevents gnats from laying eggs in it. When laid on horticultural shade cloth, paper, cardboard, and coir covers, the eggs will shrivel. A thick layer of perlite or diatomaceous earth serves the same purpose.

Trap 'Em

Yellow sticky cards are an excellent way to monitor for an infestation before one really begins, but not an effective method of control. Place the cards horizontally near the soil surface rather than hanging them vertically above the plants. Leaves can get stuck on them and they are likely to stick to the gardener when caring for the plants.

Larvae are often seen wiggling in the drain water.

Blow Them Away

Anecdotally, a 1,500-watt hairdryer aimed at flying adults kills them. While not feasible for large areas, it can be effective in small growing spaces, and head off an epidemic.

BIOLOGICAL CONTROLS

Many organisms are used to control the fungus gnat population biologically. Options include bacterial solutions, predator nematodes, a nocturnal soil mite, and a parasitic wasp. All of these are benign to humans and pets.

Bacteria: *Bacillus thuringiensis var. Israelensis (Bt-i)*

The *israelensis* strain of *B. thuringiensis* (Bt-i) is marketed as Vectobac, Gnatrol, Mosquito Attack, Skeetal, Teknar, and Bactimos. These products are available in both liquid and granular formulations. They are applied to the surface of the soil or rock wool. Bt-i is compatible with some other biocontrol agents, such as the beneficial nematode *Steinernema sp.,* and the soil mite *Hypoaspis miles.* Bt-i does not reproduce in pest populations, so it must be reapplied every week to 10 days.

Beetles: Rove Beetle *(Atheta coriaria)*

The rove beetle is a soil-dwelling insect that reproduces easily as long as prey is abundant. For a more effective treatment combine *A. coriaria* along with *H. miles* to increase effectiveness.

Flies: *Coenosia attenuate*

The hunter fly or tiger fly controls fungus gnat maggots and adults.

Fungus: *Beauveria bassiana*

B. bassiana is a naturally occurring fungus found throughout the world. It works as a contact control, affecting many types of pests, including fungus gnats. It also affects beneficial insects, a factor that should be taken into account by anyone contemplating using the fungus. It does best in high humidity and temperatures of 70 to 86 °F.

Nematodes: *Steinernema feltiae*

These beneficial nematodes invade the fungus gnat maggots by releasing bacteria that liquefy the host's bodily tissues. Beneficial nematodes can be used in all growing mediums. They are very effective.

Predatory mite: *Hypoaspis miles*

This beneficial soil mite does its work at night preying on fungus gnat larvae in the soil. *H. miles* is commercially available in 1-quart shaker bottles. *H. miles* is most effective when used early in a fungus gnat infestation, when the population is still moderate. Use yellow sticky cards placed one per square meter to gauge the usefulness of this treatment. When fewer than 20 fungus gnats per week are collected per card, *H. miles* is an effective treatment. If the card indicates a more dense fungus gnat population, *H. miles* may need help. It can be used in combination with other solutions such as rove beetles (*Atheta coriaria*).

Parasitic wasps: *Synacra pauperi*

This parasitic wasp lays its eggs inside fungus gnat larvae. These wasps have been used successfully in Swedish greenhouses. The University of Illinois was able to eliminate a chemical spray program for their Conservatory and Plant Collection by using *Synacra pauperi* exclusively for fungus gnat control. This wasp is not available commercially but is a natural parasitoid of fungus gnats and often follows their population.

PESTICIDES & FUNGICIDES
Cinnamon Oil & Tea

Cinnamon serves as a repellant for multiple insects including fungus gnats. It also works well in combination with other herbal or plant oils that function as repellants, including coriander oil, neem oil, sesame oil, and blended herbal oils or horticultural oils such as Ed Rosenthal's Zero Tolerance, which can be used at half-strength to kill fungus gnat larvae in the planting medium.

Peppermint Tea

Strong peppermint tea kills fungus gnat larvae.

This treatment may have to be repeated to get the larvae as they hatch. The tea is harmless to plants and, of course, does not harm pets and humans.

Pyrethrum

Pyrethrum sprays kill fungus gnats flitting around the garden and can be used as a drench to kill larvae in the soil.

SUMMARY
Fungus Gnat Controls

Look in Part 2: Controls for more detailed explanations of the solutions listed below. Part 2 is divided into four sections, each representing different strategies.

Preferred methods are marked with an ∗ asterisk.

Preventive Measures

Fungus gnats are often an indication of overwatering or poor housekeeping. Allow the media to dry more between waterings, and remove any garden debris promptly. Soil barriers prevent larvae from migrating to the planting media, killing them. Nematodes, *B. bassiana*, Bt-i, and herbal oils can be used to prevent outbreaks.

Barriers & Physical Controls see *Controls: Section 1*

Air-intake filters
Diatomaceous earth
Horticultural fabric
∗ Insect netting
Sticky cards
Vacuuming

Pesticides see *Controls: Section 2*

∗ Cinnamon:
cinnamon oil and tea
Coriander oil
D-limonene
∗ Herbal oils:
horticultural oils
Hydrogen peroxide
Neem oil
Peppermint tea
Pyrethrum
Sesame oil

Biological Controls see *Controls: Section 3*

Bacteria:
∗ *Bacillus thuringiensis* (Bt-i)
Beetles:
rove beetles (*Atheta coriaria*)
Flies:
Coenosia attenuate
Fungi:
∗ *Beauveria bassiana*
∗ Nematodes:
Steinernema carpocapsae
Steinernema feltiae
Predatory mite:
∗ *Hypoaspis miles*
Parasitic wasp:
∗ *Synacra pauperi*

Outdoor Strategies

Remove any standing water or moist piles of plant refuse.

Gophers

Gophers are tenacious and resourceful garden pests that frustrate many gardeners. Take poor Carl Spackler and his ongoing battle with a stubborn gopher in the cult classic *Caddyshack*. If you've got gophers, you've got trouble.

"Gopher" is a much-misused term that often refers to several unrelated animals including ground squirrels and moles. True gophers are known as pocket gophers, so named for the fur-lined external pouches on their cheeks that they fill with food or nesting material. These pouches open to the outside of the body and can be turned inside out to empty or to clean.

These burrowing rodents live almost exclusively in an extensive system of underground tunnels that can cover 200 to 2,000 square feet, resulting in the movement of more than 4 tons of soil! From this labyrinth of underground passageways, they graze on plant roots, fleshy underground plant parts, irrigation pipes, and even underground utility lines. They live their lives mostly unseen, except for the rare occasion when they are spotted shoving dirt out of newly excavated tunnels.

Excellently suited for a life of digging, gophers have powerful front legs with large, clawed paws they use to move pounds of soil as they create their underground world. Their lips can close behind their four large, prominent front teeth, leaving the teeth always exposed but keeping dirt out of their mouths as they chomp their way through dirt and tree roots. As is true with other rodents, gophers' teeth grow as much as 14 inches per year and they must chew and gnaw continuously to keep the length under control.

Gophers have tiny eyes and poor eyesight. Their ears are small as well, and lay flat against their head to keep out dirt. Their lack of vision is compensated by the enhanced sensitiv-

At a glance

Symptoms
Outdoor signs of tunneling or horseshoe-shaped mounds of dirt with off-center holes; positively identifying gophers; signs of weakened plant roots; wilting plants.

Gopher emerging from hole.

ity of their long whiskers, or vibrissae, which are sensitive to movement and useful for navigating underground. Long, naked tails help them scurry backwards as fast as frontwards. Depending on the species, gophers range from 7 to 12 inches long. Their fur is dense and pliable, ranging in color from rich brown to yellowish to gray.

There are five different types of gophers, based mainly on where they are found:

Types of Gophers			
Genus	Gopher	Location	Likely To Be Encountered
Geomys	Eastern (Plains) pocket gophers	Southeast United States; east of the Sierra Nevada	Yes
Orthogeomys	Taltuzas	Mexico, Central America, and Columbia	No
Cratogeomys	Yellow-faced pocket gophers	West Texas, east New Mexico, Kansas, Mexico	Limited range
Pappogeomys	Mexican pocket gophers	Mexico	No
Thomomys	Western pocket gophers	Western United States	Yes

Among these types are about 34 different species of gophers on the North and South American continents from Canada to Colombia. They are divided into five genera and two tribes: the *Geomyini* tribe, with 25 species, and the *Thomomyini* tribe, with nine.

Gophers belonging to the genus *Geomys* are considered eastern pocket gophers, living from the Rocky Mountains to the Mississippi River Valley and the Gulf Coast. Those belonging to the genus *Thomomys* are western pocket gophers and range from the Rocky Mountains to the Pacific Ocean and from southern Canada to Mexico. *Cratogeomys castanops*, or the Mexican pocket gopher, is found from the southwestern United States to central Mexico. There are other genera of gophers found in Mexico and Central America, such as the taltuzas and yellow and cinnamon pocket gophers. Gopher ranges rarely overlap between species.

The presence of gophers can be identified by mounds of freshly excavated dirt, usually horseshoe or crescent shaped, with an entry hole off to one side that is plugged with dirt. Mole holes are sometimes mistaken for gopher holes, but mole holes are usually circular or volcano-like with an opening in the center.

LIFE UNDERGROUND

Gophers are the hermits of the rodent world and are noted for their antisocial lifestyle. They live alone in their elaborate underground tunnel system, interacting with other gophers only during mating season when the male gopher visits the female in her tunnel system. Occasional non-reproductive meetings with other gophers do not go well, with excessive hissing, squealing, and teeth-chattering, and even an occasional all-out brawl.

Some populations are crepuscular—they are active mainly at dusk and dawn. Other populations are nocturnal, active only at night.

Gopher tunnels are generally about 2 ½ to 3 ½ inches in diameter, 6 to 12 inches deep, and generally run parallel to the surface. Most of their tunnel system is for moving about undetected and seeking food.

Gophers dig their tunnels by clawing and chewing their way through the soil with their powerful front feet and teeth, pushing the loose soil backwards with their back feet. When enough soil has accumulated in the tunnel, the gopher uses its front claws and chest to shove the soil to the surface through short, sloping tunnels dug just for this purpose, thereby creating the classic fan-shaped soil mounds. Once a gopher has excavated all the dirt desired in the area of these tunnels, they will be plugged with dirt as the gopher moves on to dig new tunnels.

Gophers do not hibernate. Instead, they store food for the cold months. They construct numerous food storage chambers deep underground in their tunnel system. Then they carry plant material to store in these chambers for eating later in the season by stuffing their cheek pouches.

Gophers do most of their tunnel building in the spring and fall, when the soil is moist and easy to move. In irrigated areas where the soil remains soft, they dig year round. Gophers can dig 200 to 300 feet in a single night. Understandably, gophers also dig more profusely in sandy soils rather than hard soils. In areas with deep snow cover in winter, gophers tunnel through the snow to feed on tree bark. Once they have had their fill, they plug these snow tunnels with dirt and retreat back to their underground home. When the snow melts in spring, long cores of soil remain behind, telltale evidence that gophers have been there.

Food & Feeding

Gophers are herbivores and eat up to 60 percent of their body weight each day. They ingest nearly all roots that they encounter, including fleshy below-ground plant parts such as bulbs and rhizomes, and roots of other plants, including cannabis. They prefer to travel and feed on plant parts accessible from inside their tunnels, but they sometimes emerge a few body lengths from their tunnels to gather food above ground to take back to their tunnels. Occasionally they even drag whole plants back into their burrows. Burrows used in this manner are called feed holes, and the gophers construct them without the characteristic mounds of dirt. They are identifiable because the opening to the hole is surrounded by close-clipped vegetation.

Gophers do not need a readily available source of water because they get most of the moisture they need from their food. However, in arid climates, they eat cactus as a source of water.

Gopher Sex

Gophers reach sexual maturity when they are a year old. They live as long as three years. They mate in early spring through early summer. Male gophers dig lateral tunnels, hoping to intersect the tunnel of a female. The romance is only a one-night thing. They quickly part company,

gopher hole

molehill

each returning to their solitary lifestyle.

The female makes her nest chamber about 10 inches in diameter and as deep as 6 feet underground — a handy depth in areas where the ground freezes and predators are avid diggers.

The female gopher lines the nest with dried plant material in preparation for the birth. After a gestation period of about 18 to 19 days, she has a litter of three to 13 pups. The young are born with their eyes and their cheek pockets closed. The pouches open at 24 days and the eyes at about 26 days. The pups mature quickly, in about five to six weeks. Once mature, they simply wander away to find their own territory and begin their own system of tunnels.

In irrigated areas where the soil is soft and the living is easy, they can produce up to three litters per year and population density increases. In irrigated fields of alfalfa, one of their favorite foods, gopher densities may reach 60 gophers per acre, each one with its own honeycomb of tunnels. In drier areas where life is harder, females give birth to one litter per year consisting of three to seven young.

Mortality is high among gophers and in some populations, only 6 to 12 percent of the young survive to reproduce the next year. They are preyed upon by a large variety of predators including coyotes, hawks, owls, foxes, snakes, skunks, badgers, weasels, and domestic cats and dogs. In areas where there is a deep snow pack, spring melting can flood the burrows and drown the gophers. Females live nearly twice as long as males, attributable mostly to the aggressive and pugnacious nature of the males during breeding periods. Males also continue to grow throughout their lives, expending valuable energy in the process. Females cease growing when they reach maturity.

Damage

Gophers wreak havoc on any crop they find particularly tasty. They nibble away at the roots that extend into their tunnels, returning to nibble some more when the plants produce new roots. Eventually, the root system becomes so atrophied that it can no longer absorb enough water and nutrients to sustain the canopy. The plant withers and dies or else hangs on and grows stunted.

Anything that lies in the path of a digging gopher is fair game for nibbling, even if it's not an ideal snack. They can do extensive damage

Northern pocket gopher: close-up of pouches.

to irrigation piping, underground utility cables, and water pipes. Gopher mounds and holes are dangerous if stepped in by an animal or human or small-wheeled vehicle.

Benefits of Gophers

The very idea of gophers as beneficial organisms is laughable, even ridiculous, to gardeners, especially as a garden's plants disappear. But gophers do some good—maybe not enough to outweigh their damage—but some good, nonetheless.

First, they're terrific soil conditioners. The feces and plant material they leave behind in their tunnels decays and enriches the soil. The presence of decaying plant material in their tunnels encourages earthworms and other beneficial insects to move in. Additionally, their tunneling aerates the soil, which increases the amount of water that can percolate through the soil, reducing soil compaction. Through their tunnel excavations, they bring subsoils to the surface, making minerals available to shallow roots.

Finding Gophers

Gopher damage is often confused with moles or voles, two other underground tunneling creatures. However, an educated gardener can tell the difference. Moles are members of the *Talpidae* family. These underground dwellers are significantly smaller than gophers, about 1 to 3 inches in length. They also feed on plants roots and the fleshy parts of plants, but about 90 percent of their diet consists of earthworms, beetles, fly larvae, slugs, and other invertebrates that fall into their tunnels.

Voles are rodents and members of the *Muridae* family. They more closely resemble mice and rats in their lifestyles and habits than gophers. They are about 4 to 4 ½ inches in length. Some species, but not all, are avid tunnelers. They are herbivores and forage for food both at night and during the day. Some voles are agricultural pests, such as the American pine vole that burrows around the bases of apple trees in the winter where it chews on and girdles the roots and stems. Many species of voles live in areas of little agricultural value and are generally not considered to be pests of any great concern.

Correctly identifying the source of the problem is extremely important before initiating control measures. The chart below offers a comparison of gophers versus moles.

Differences between moles and pocket gophers	
Moles	**Pocket Gophers**
Minute eyes are often not visible.	Small eyes are clearly visible.
Muzzle is long and tapering.	Muzzle is rounded.
The many small teeth are not apparent.	Orange, chisel-like pairs of upper and lower incisors are apparent.
Mounds are round when viewed from above.	Mounds are crescent or heart-shaped when viewed from above.
Soil plug is in the middle of mound and may not be distinct.	Soil plug is in the middle of the V shape or off to the side of the mound and may leave a visible depression.
Tunnels are often just beneath the surface, leaving a raised ridge.	No tunnels are visible from above ground.

CONTROLS

Obviously, gophers are really only a concern for the outdoor gardener. Those faced with a population of gophers should think of them as tenacious when arming themselves. Because they live most of their lives underground, gophers are hard to see, much less eradicate. They are resourceful and notorious at thwarting efforts to dislodge them from their nice, cozy homes.

Prevention

If you know there are gopher problems in your yard or outdoor grow area before you plant the garden, raised beds may be the best way to go. To create more protection for the roots of raised-bed garden plants, secure ½-inch hardware cloth to the bottom of the bed, then fill and plant.

Enclose small garden plots using hardware cloth beneath the soil and up the sides of the area to be planted. Individual plants can be protected using hardware cloth to line planting holes or protecting the bottom of raised beds. To protect watering pipes or utility lines, lay 6 to 8 inches of 1-inch diameter gravel around the lines or wrap them in wire mesh.

To stop gophers from digging into a raised bed, place a layer of fencing at the bottom of the bed that extends out from the perimeter by about 6 inches. If you are using planting holes in gopher country, line the entire hole with wire. This provides total protection from gopher predation.

Traps

Trapping is probably the best, most reliable, way of controlling gophers. But, before you can trap them, you have to find them. First, locate a dirt mound with dark, moist soil as an indication of recent digging. On the "unfanned" side of the crescent-shaped mound, insert a metal or wooden rod into the soil about 8 to 10 inches away from the hole and about 6 to 12 inches deep. If there is a tunnel below the rod will suddenly drop down 2 to 3 inches. Continue to probe until you find a larger excavated area beneath the surface. This is the gopher's main burrow. Disturbing the lateral tunnels by probing causes the gopher to abandon and seal them off, so trying to trap them in those areas is usually fruitless.

CHOOSING A TRAP

Several types of gopher traps are available. The most common of these is the two-pronged pincher trap, such as the Victor Black Box, Macabee, Gophinator, or Cinch. With these traps, the gopher triggers the trap by pushing against a flat, vertical plate. Another type, the choker-style box trap, is easier to set and use for those unfamiliar with gopher traps. The box-type trap requires a larger surface opening, but works well in smaller gopher tunnels.

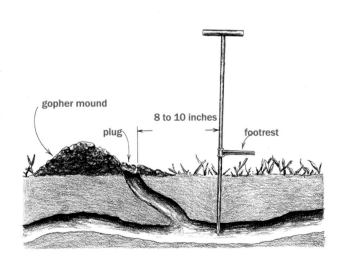

gopher mound

plug

8 to 10 inches

footrest

SETTING THE TRAPS

Once you have located the main burrow and the lateral tunnels leading away from it, place your traps in tunnels on either side of the main burrow, facing in opposite directions so the gophers will be caught coming to the main burrow from either direction. Be sure to wear gloves to keep your scent from clinging to the traps. Secure the traps by attaching a length of cord, twine, or chain to the trap and to a wooden stake driven into the ground so you won't have to go looking for the trap a gopher dragged deeper into the tunnel system.

Gopher traps don't have to be baited but trappers sometimes place carrots, apples, lettuce, alfalfa greens, or peanut butter at the back of the trap so the gopher has to trigger the trap before reaching the food.

Cover the hole with a board wider than the hole you dug and then cover the edges with dirt to shut out all light because light at the end of the tunnels spooks the subterraneans. A few gopher experts contend that leaving the hole open lures them into the trap because of their urge to plug holes in their tunnel system. Leaving the hole open does make checking and resetting traps easier and less time-consuming.

Check the traps every other day. If no gophers are caught after two or three inspections, probe around until you find another tunnel. Repeat the process until you (hopefully) catch the gopher.

Barriers

Barriers are one of the most effective ways of controlling gophers. They can't eat what they can't get to. Barriers can be work-intensive to install and are not foolproof. Intelligent gophers quickly learn to dig deeper to reach those tasty roots.

Gophers live underground, so the fence has to block them below ground, not above. For gopher problems around a garden of established plants, bury galvanized ½-inch hardware cloth or ¾-inch poultry fencing 2 feet deep all around the perimeter of your planting area with about 6 inches to a foot of additional wire above the ground bent at a 45-degree angle away from the plants. This should deter gophers from digging to the plant roots and also prevent them from climbing out of their tunnel to simply step over the wire.

Exposure

Cut the vegetation close around gopher

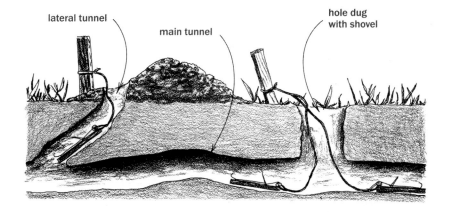

lateral tunnel

main tunnel

hole dug with shovel

holes to force the pests to expose themselves to predators as they search for food.

Flooding

On irrigated farms, area flooding is sometimes used to provide gopher control. Entire fields are briefly flooded. Water sits on the land long enough to percolate down into the gopher burrows, drowning them or driving them out. Flooding can also kill plant pathogenic fungi, nematodes, and soil insects. Home gardeners should not assume that they can chase gophers away arming themselves with a water hose. When one part of the tunnel is flooded, the gopher moves to another section that is still dry then dams the tunnel so the water won't reach. Occasionally a gopher is driven out of its burrow. In order to successfully flood the tunnel system, the gardener must understand its size and be able to accomplish a thorough flooding.

Fumigation

Fumigation may work best as a part of a larger campaign against gophers. Pest control experts and researchers debate whether fumigation can effectively control gophers. CO_2 paper cartridges filled with charcoal and potassium nitrate are the most common fumigants available. CO_2 can also be delivered into the gopher tunnels either by placing a hose connected to a CO_2 tank in the gopher's tunnel, or by dropping 8 to 16 ounces of dry ice into the tunnel opening.

Smoke gas cartridges are also available commercially, literally smoking rats and gophers out of their holes. They contain carbon monoxide and, when ignited, fill the burrows with gas that renders the animal unconscious. Death from as-phyxiation soon follows. Use according to label instructions.

The problem with fumigation methods is that gophers plug up parts of their tunnel system that they deem at risk, or when they sense a predator attack. So, at the first wisp of gas, they actively plug the tunnel and move into a safer location.

Predation

A number of predators, including owls and snakes, are quite accomplished at picking off gophers as they venture from their tunnels. Unfortunately, natural predation won't eradicate a gopher population because predators move onto more fertile hunting ground when the gophers become scarcer and harder to catch. Although this method is described as "animals as pest control," it is completely impractical to try to harness toward pest management.

Irritating the Gophers

There are numerous other espoused methods for repelling, frightening, or killing gophers. However, these methods probably do little more than irritate the gophers but don't eradicate them. Gophers are tenacious and unlikely to be budged.

Repellants

Repellents and scare tactics rarely work long-term for gophers. Their use might slow up the destruction of the cannabis patch a little, but neither is likely to send the little pests packing. Just as with trying to bug them into leaving, gophers are hard to scare.

They are attenuated to human sounds: lawn mowers, tractors, lawn edgers, leaf blowers, combines, and sprinklers. They quickly determine which sounds are a threat and which are benign. With their system of tunnels that maze under your garden, trying to frighten gophers isn't likely to chase them away from an easy meal. Scare devices such as wind-driven pinwheels, vibrating sticks, and ultrasonic devices are available. Don't be tricked into using them because you will almost certainly be disappointed.

GOPHER CONTROLS

Preventive Measures

Gophers travel through tunnels. You can prevent damage using mesh wire, or poultry fencing in the layout of the garden. First step is to make sure the animal is not a mole.

Solutions are specific to gopher behavior. Preferred methods are marked with an * asterisk:

* Traps
* Gopher fence

Flooding
Fumigation

Predation:
cats, owls, snakes

Grasshoppers

Grasshopper on cannabis.

More than 11,000 species of grasshoppers live on the North American continent and new species continue to be identified. They are very adaptable and are found in nearly every habitat on earth except the North and South Poles.

Grasshoppers are members of the order Orthoptera which includes two families: Acrididae and Tettigoniidae. The family Acrididae is part of the suborder Caelifera or "short-horned grasshoppers." They have short, thick antennae that are not longer than their body. They chirp by rubbing their legs against an organ on their abdomen. The Acrididae are often referred to as "true grasshoppers" and include the dreaded locusts among their numbers.

The Tettigoniidae family is a member of the suborder Ensifera and are called the "long-horned grasshopper." Logically, their antennae are longer than their bodies and slender, and include groups such as crickets and katydids. They make their chirping by rubbing their legs together.

IDENTIFICATION

Grasshoppers ravage crops with powerful chewing jaws and bound away to great distances when disturbed. Signs of grasshopper damage are large irregular holes chewed in leaves. However, their damage can be confused with several other insects including caterpillars and flea beetles, so spotting them at work is the best way to determine the culprit.

The grasshopper is a distinctive-looking insect whose profile and method of travel allow most people to easily identify it, at least generally if not specifically. Grasshoppers have four large wings, chewing mouthparts, and powerful hind legs. They range in color from bright green to grayish brown. The front pair of wings, or tegmina, are leathery and are mostly for protection of the hind membranous pair, which can be very colorful, depending on the species. Like all other insects, grasshoppers have a hard exoskeleton and three body segments—head, thorax, and abdomen. They breathe through a series of small holes in their abdomen called spiracles.

Grasshopper species vary in size from ½ inch to over 4 inches, depending on the species. They have three pairs of legs that are all used for walking. The hind pair is thick and powerful and is used for jumping from place to place or for launching them into flight. Grasshoppers can jump 20 times the length of their body in a single bound and launch themselves at the velocity of 3 meters per second, about 6 miles an hour.

Grasshoppers specialize in biting and chewing. Their jaws, called pinchers or mandibles, cut and tear off their food. Two large lips, an upper called a labrum and a lower called the labium, maneuver the food into their mouths where grinding jaws are lined with small sharp teeth that macerate the food in a side-to-side motion. A second set of jaws behind the mandibles, called maxillae, push the food down the throat.

Grasshoppers have two large compound eyes on each side of their head. Each eye is made up of thousands of tiny eyes that work together to complete a picture in the grasshopper's brain. They can look forward, backwards, and sideways at the same time. They also have three simple eyes (ocelli), one above each antennae and one in the center of their face. Simple eyes detect changes in light intensity.

THE GRASSHOPPER LIFESTYLE

Grasshoppers are most likely to thrive in semi-arid areas with less than 25 inches of rainfall a year. Grasshopper populations fluctuate from year to year based mostly on rainfall and temperature.

Grasshopper behavior is closely linked to immediate environment. Temperature, day length, rain, and wind all play a part in how a species of grasshopper behaves. Behavior varies among grasshopper species, giving each one special survival abilities. Their adaptability as well as their finely honed ability to detect and evade predators enable grasshoppers to survive, tolerate, and adapt to a variety of environments.

At night, insects, which are cold-blooded, become lethargic as the temperature drops. Sensing the cooling temperature, grasshoppers seek shelter. Species pass the cool night hours in different ways. Some seek shelter under leaves, grasses, or leaf litter while others squat on bare ground and await the sunrise without taking any special actions. Others climb onto plants and grasses and cling there until the light returns.

The grasshopper's day begins a little after dawn. As the temperature increases it becomes more active, and crawls out of its hiding place to bask in the sun. Ground-dwelling grasshoppers crawl into a sunny spot on the ground and those in plant foliage creep out onto a sunny leaf.

To gain heat grasshoppers turn the long, narrow side of their body perpendicular to the sun and drop the muscular back leg on that side so the rising sun's heat rays fall on their abdomen. Grasshoppers remain still and quiet while they soak up the sun. After an hour or two, they are sufficiently warmed to begin their day. Occasionally they will preen. Sometimes they turn around to let the sun warm their other side. In the late afternoon, they bask a second time before the sun sets and the night cools.

They spend their time feeding, seeking out mates, or simply moving from place to place. In late spring and early summer, grasshoppers forage for food all day. Unseasonably cool or rainy weather disrupts this cycle and they creep back to their hiding places to wait for the sun and its warmth.

A feeding community of grasshoppers is a study in cooperation. Some species climb their host plants to feed on the leaves, buds, or stems. Others simply chew away at the base of a grass leaf until the leaf topples over. They can then feed on the plant on the ground at their leisure, safer from predators than bobbing about on waving plants. Ground-dwelling grasshoppers forage for seeds, leaves felled by other grasshoppers, and the occasional dead arthropod. In their natural grassland habitat, evidence of their existence is often nearly undetectable because food dropped by one member of the community is consumed by another.

Despite common belief, grasshoppers are pretty choosy about their food. Sensory organs called gustatory sensilla located in the tips of their antennae allow them to detect the attractant or repellent properties of plant chemicals in potential host plants. They lower their antennae to the surface of the plant leaf and drum or tap on the leaf surface until they are satisfied this is a tasty plant or one they want to avoid. If still in doubt, the grasshopper may take a bite of the leaf to help it decide.

Phagostimulants are important plant nutrients such as sugars, phospholipids, amino acids, and vitamins that can be tasted by animals and stimulate their feeding impulse, so named from the Greek *phagein* meaning to eat. Detection of

American grasshopper.

these nutrients stimulates grasshoppers to select one plant over another or even one leaf over another. Grasshoppers prefer new young growth, rich in sugar, rather than old yellowing growth with few sugar reserves.

Grasshoppers have developed various methods of dealing with peak daily summer temperatures. When the ground temperatures reach 120 °F grasshoppers climb to various heights on plants and grasses. Members of some species climb up an inch or 2 while others climb as high as possible. In reverse of their morning routine, they turn their face toward the sun to minimize their body's exposure to the sun's rays.

Ground-dwelling species rise up on their back legs, or stilt, to raise the bulk of their body away from the hot ground and allow air circulation around their abdomen. As the air temperature continues to rise, they move into the shade until temperatures drop with the waning of the day.

Ironically, the majority of a grasshopper's day, as much as 80 percent, is spent nearly motionless. Very little time is actually spent feeding, mating, seeking mates, or moving from place to place.

If disturbed, grasshoppers bound or fly away in what appears to be a mad dash. The stimuli that it takes to make a grasshopper move are another curiosity. A stick thrust into the ground near a grasshopper is usually ignored. Large beetles or other larger insects crawling by are ignored as well unless they touch the grasshopper. Plant-feeding grasshoppers move to the underside of their leaf or branch, move deeper into plant cover, or drop to the ground. Grasshoppers are most disturbed when a shadow passes across the surfaces of their compound eyes.

Communication

The slow, rasping chirp of grasshoppers on a hot summer day is permanently embedded in many people's memories of summer. No summer afternoon or evening is complete without the chirping of grasshoppers and their relatives. Grasshoppers communicate with these acoustic signals as well as using visual communication to seek mates, defend food or territories, and to ward off unwanted sexual advances. Each species has its own unique language.

The method grasshoppers use to produce their chirping helps scientists distinguish between species. They don't have vocal chords, but use specialized ridges on the inside of their hind legs known as stridulatory pegs that they rub against their leathery forewings, called tegmina. These rubbings, done in various ways and for various durations, create sounds that other grasshoppers "hear" using specialized auditory organs on the first segment of their abdomen. In most species, only the male grasshoppers sing.

The sound we hear when a disturbed population of grasshoppers leap into the air and fly away is called creptitation. This sound is created by the grasshoppers rapidly snapping their hindwings in flight and may be a signal to their companions that danger is imminent.

Life Cycle

Grasshopper females lay their eggs in pods they deposit into narrow tubular holes in the ground that they dig with their ovipositor, a specialized structure on the end of their abdomen. The number of eggs laid varies greatly with grasshopper species. The eggs begin development as soon as they are laid. Most species of grasshopper eggs reach a point in their development when they enter diapause, a phase during which development slows or stops. Diapause is triggered by low temperatures maintained for a length of time.

With the advent of spring, the soil warms and the eggs resume their development and hatch as nymphs closely resembling their parents. Grasshoppers grow from egg to adult through an incomplete metamorphosis. When they hatch, grasshoppers look like miniature adults. An entire clutch of grasshopper nymphs wriggles out of their soil nursery in mass, one tiny grasshopper after the other struggling to the surface of the soil. Each nymph emerges contained in an embryonic membrane called a serosa. They quickly shed this protective sack. At this point, they are most vulnerable to predation by foraging ants.

Once free of their embryonic wrap, the nymphs hop away, feed, and try to avoid predators. While closely resembling their parents in appearance, the tiny nymphs cannot fly but have wing pads where wings will form later. The nymphs of most grasshopper species molt, or shed their skin, four to six times before reaching adulthood. After the last molt the now-adult grasshoppers have fully functional wings and can fly. The females have a one-to-two-week period during which they gain weight and mature before they are ready to mate and lay eggs.

Once a female has mated with a male, she immediately begins to lay eggs. With other males in attendance, the female digs a hole in the soil with her ovipositor and lays her first batch of eggs. As soon as she retracts her ovipositor, another male steps up to mate with her. A female grasshopper continues to lay eggs for the rest of her short life.

Locusts

Locusts are grasshoppers whose morphology, color, and behavior changes at high-population densities. When grasshoppers are crowded to the point of bumping into each other, they undergo physical changes and become locusts. They are characterized by their behavior and do not belong to any one genus or species. Generally, grasshoppers classified as locusts are short-horned grasshoppers or members of the Acrididae family that have swarming phases of their life cycle. Most grasshoppers are solitary insects, passing their lives alone except during mating. Locusts, however, suddenly shift their behavior and group into massive numbers of insects and begin to migrate into new territory looking for food.

While single grasshoppers are only a curiosity to man and mildly irritating as they munch on garden crops, locusts have been feared far into the past, landing them significance in the Bible and the Jewish Torah. What triggers this swarming behavior is not completely known, but some species swarm in response to spring rains or warming temperatures.

SPECIFIC GRASSHOPPERS ON CANNABIS
Two-striped Grasshopper (Melanoplus bivittatus)

The two-striped grasshopper is named because of two light-colored stripes that run the length of the body and then come together posteriorly to form a triangle. They are short-horned grasshoppers and are among the earliest to emerge in the spring. Females are large, averaging about 1 ½ inches in length. The males are smaller at about 1 inch. Two-striped grasshoppers are most recognizable by their distinctive light stripes on the head, thorax, and forward edge of the forewings.

This grasshopper bears a short "spur" that protrudes outward from between its front pair of legs. They overwinter underground as eggs that are 60 to 80 percent developed. In southern areas, nymphs emerge in June and in more northern climes, individuals appear by July. Each female may lay one or two batches of eggs. Their distribution ranges from southeastern Canada to Mexico. They frequent grass pastures and clover fields. These grasshoppers have been known to migrate in masses, stripping plants in their path.

Sprinkled Locust or Sprinkled Broad-Winged Grasshopper (Chloealtis conspersa)

The sprinkled locust's characteristic slanted face is angled from the bottom of the jaw to the top of the head. It is the only group of grasshoppers in North America in which the males, and sometimes the females, have a stridulatory file on the inner side of their hind legs. They rub this file against their leathery first pair of wings to sing or chirp.

Located from Maine westward to Alberta, Canada, south to Colorado, and eastward to North Carolina, it prefers grassy thickets along-side streams, fields, or pastureland.

Female color varies from a dull yellow to a dark brown, and the front wings are speckled with small black spots. Males are light brown with a black bar across the thorax. They do not have the speckling on their front wings. Females are about 1 inch long. Males are a bit smaller, averaging around ¾ inches in length. Both males and females have red-to-yellow hind legs.

Eggs laid in the fall overwinter underground in masses of 15 to 50 eggs encased in a hard, gummy substance.

Clearwinged Grasshopper (*Camnula pellucida*)

Clearwinged grasshoppers are short-horned and are members of the "band-winged" grasshopper subfamily, but this species has no bands on its wings. Its range is from Alaska south to Mexico, northeastward to the Dakotas, then east to Nova Scotia.

Females begin their life as a black nymph with white-to-tan markings. Adults are dull brownish-gray with black markings. When they become sexually active, they turn a bright, se-ductive yellow. Males average about ¾ inches in length and females are about 1 inch long.

Clearwinged grasshoppers overwinter as eggs laid in masses of 10 to 30 underground in bare places such as tilled land and along road-sides. They are one of the earliest emerging grass-hoppers and the nymphs sometimes migrate in swarms. When population densities are greater than 25 individuals per square meter, they dev-astate rangeland.

Melanoplus bivittatus: two-striped grasshopper.

Chloealtis conspersa: sprinkled locust or sprinkled broad-winged grasshopper.

Camnula pellucida: clearwinged grasshopper.

Grasshopper Cousins: Crickets & Cockroaches

Included in the order Orthoptera are crickets and cockroaches. Both of these insects attack marijuana. Their damage appears as holes chewed in the leaves that are often confused with cutworm, damping off, or root damage similar to that caused by root maggots or root grubs. However, these are probably incidental occurrences and neither insect poses a great risk to a cannabis patch.

CONTROLS

Grasshoppers can become pests in marijuana gardens, but they are not motivated to seek out marijuana as their primary diet. If the population isn't so large that it is stressed to find food sources, simple barrier methods such as mulch can deter grasshoppers from the garden. Given their size, they are easy to keep out of indoor gardens. Simply make sure there are no sizeable leaks through which they can enter the garden.

If there's food pressure or the grasshopper species is interested in cannabis, the passive barrier forms of repellant may need to be bolstered by more active methods. However, the repellant has to be selected carefully. Some remedies such as vegetable oils and garlic-based applications attract rather than repel them. However, a hot pepper (or capsaicin) spray made from hot peppers, pure soap, and water repels some species.

Grasshoppers avoid cilantro. It may have some repellent qualities when planted all over the garden as a companion plant. It may also work as a spray ingredient.

PESTICIDES
Neem Oil

The oil pressed from the seed of the neem tree (*Azadirachta indica*) has both insecticidal and fungicidal properties. It is safe for mammals, including humans, and degrades quickly once it is

Grasshopper laying eggs on marijuana leaf.

exposed to the environment. It is used against many insects so it is a good multipurpose treatment or prophylactic in the garden. Neem oil can be applied up to three weeks before harvest.

Pyrethrum

Pyrethrum is a natural contact insecticide derived from the chrysanthemum family. Like neem, it works against most of the major insect pests known to marijuana gardens, but unlike neem, it can be toxic to beneficial insects, fish, and reptiles. Pyrethrum is available in powder and spray form, and is appropriate for hobby-sized gardens or larger applications. Gardeners should limit their own exposure in indoor garden applications, and should make sure they are free from pyrethrum allergies.

Leaf damage.

BIOLOGICAL CONTROLS

Some biological controls are effective against grasshoppers.

Fungi

BEAUVERIA BASSIANA

This naturally occurring fungus is used against many insects, including thrips, whiteflies, termites, aphids, some beetles, and grasshoppers. It works by penetrating the host body then proliferating inside until the insect dies. The fungus then emerges from the insect body and produces more spores to establish the fungus in the environment.

ENTOMOPHTHORA GRYLLI

This fungus is native to North America, Europe, and Africa, and infests crickets and grasshoppers. Spores that touch the insect germinate and

penetrate the exoskeleton. The grasshoppers die seven to 10 days after infection.

Protozoa

NOSEMA LOCUSTAE

This microsporidian protozoa is a natural parasite of grasshoppers. It takes time for this organism to establish itself, so it is not a quick fix. However, if the problem is sustained or the grasshopper population is high in your area, *N. locustae* can reduce grasshopper numbers. *N. locustae* is available commercially, but it is intended for large acreages with huge grasshopper problems, usually on rangeland. It is not effective for backyard use.

OUTDOORS

In outdoor gardens, clearing the land surrounding the garden is a deterrent. By eliminating sources of food and shelter near the garden and losing protection from predators, grasshoppers may be reluctant to venture into the area. Domesticated fowl such as chickens and turkeys, and wild birds using bird feeders also discourage grasshoppers from entering the area, since these birds enjoy grasshopper snacks. Insect netting or screening is very effective for keeping them from the plants. Tilling the garden in late summer reduces the chance of repeat infestations of future crops.

SUMMARY

Grasshopper Controls

Look in Part 2: Controls for more detailed explanations of the solutions listed below. Part 2 is divided into four sections, each representing different strategies.

Preferred methods are marked with an ✱ asterisk.

Preventive Measures

Use netting to protect plants if grasshoppers are expected to become a problem. Chickens or other domestic fowl help control large insects like grasshoppers.

Barriers & Physical Controls see *Controls: Section 1*

    Air-intake filters

✱ Capsaicin

✱ Insect netting

    Mulch

    Pest removal

Pesticides see *Controls: Section 2*

    Insecticidal Soap

    Neem oil

✱ Pyrethrum

Biological Controls see *Controls: Section 3*

    Fungi:

✱   *Beauveria bassiana*

✱   *Entomophthora grylli*

    Protozoa:

      Nosema locustae

Outdoors see *Controls: Section 4*

    Animals as pest control: birds, fowl

    Clearing land

    Companion plants outdoors: cilantro

Gray/Brown Mold (Bud Rot & Leaf Mold)

Gray mold spores float in the air so they are always in contact with plant leaves, their possible hosts. However, they require a moist environment to thrive. The infected area first appears soggy and browned. Then the mold develops a silvery-gray, fuzzy covering composed of thousands of grape-like clusters of spores. The cool wet conditions the mold prefers occur most often in early spring and fall harvest time outdoors, or in under-ventilated and humid areas indoors. Host plants include not only cannabis, but many other fruits, flowers, and trees. Dense buds that trap water in crevices, and fruits with a high liquid content such as grapes, tomatoes, and strawberries are most susceptible.

Botrytis cinerea often develops in the tight floral clusters of cannabis where airflow is restricted. Mold in the center of a bud may not be immediately apparent from a casual look, and a nice-looking large bud may in fact be "hollow" due to mold damage. This gives rise to its common name "bud rot," although the proper name, *Botryotinia fuckeliana*, seems appropriate for this pathogenic fungus.

When the fungus kills sprouts, it is known as damping off.

Outdoors, thick buds ripening in cool, damp weather can be affected by an outbreak. Tightly packed flowers shelter the spores, providing the cool, moist, acidic environment they need to develop and spread. Wounded or diseased foliage is also a suitable host site for the mold. Indoors, poor air circulation results in still pockets where dampness collects, encouraging fungal growth.

B. fuckeliana is a parasitic fungus that can destroy a plant or buds when left unchecked. As a fungus feeding on the plant's nutrients, it needs no light to grow and prefers cool, damp,

At a glance

Symptoms
Gray-white, fuzzy, cotton-like mycelia over a grayish-brown slimy coating. It often develops in the center of floral clusters before spreading though the rest of the bud. It is also one of the fungi that causes "damping off" in sprouts.

Bud rot occurs very quickly under favorable conditions.

Botryotinia fuckeliana.

Botrytis cinerea.

Gray mold.

acidic, underlit locations. If the spore lands in a moist acidic spot it will sprout and grow strands of hyphae. As these strands fill in the area, the colony becomes easily visible to the naked eye.

The mass of hyphae strands develops into mycelia—the gray-white thread covering that is the obvious indication of a colony. The mycelia absorb nutrients from the host-plant material, destroying leaves and buds in the process. Enzymes are released that break down the plant matter into a nutrient slime that the mold can absorb. This causes the most damage to the host plant.

As long as the conditions remain favorable and there is plant-host material, it continues to spread. At the same time it grows reproductive structures that protrude above the surface of the mycelia that release airborne spores. A *B. fuckeliana* spore can germinate and mature to produce more spores in a matter of days.

A single colony releases billions of spores. Each is capable of starting a new colony. If a spore lands in a suitable environment, it will germinate and continue the cycle. Spores that land in an environment not conducive to germination remain dormant until the environment changes. They die within a few weeks in a moist environment, but in a dry environment spores can stay dormant for years.

Spores can enter the indoor garden through the ventilation system, on clothing, or on contaminated plants. Proper housekeeping can reduce the number of suitable host sites in the garden, but preventing all the spores from entering the garden is unrealistic in most situations since they are ubiquitous.

The bud is rotten inside.

Gray mold dries out the outside of a bud.

Powdery mildew and gray mold on a bud.

Controls

The best way to avoid gray mold is to not give it a suitable home to get started. Since the mold develops only in overly moist conditions, the first line of defense is controlling humidity to keep moisture levels under 50 percent. Good air circulation throughout the garden keeps humidity levels from rising to dangerous levels inside the buds.

Gray mold is likely to be most severe within the thickest part of the canopy where the air circulation is poorest. Thick outdoor plants can be spread with twine or netting to improve airflow.

Remove interior and weak branches to improve circulation. Fans assist in mold prevention by removing pockets of high-humidity air surrounding the leaves and flowers.

Don't foliar feed plants during the last 30 days of flowering because moisture may become trapped inside the buds. Foliar feed early in the day during vegetative and early flowering so moisture on leaves or buds has a chance to dry before darkness. Watering the garden at the beginning of the light period or early in the day instead of at the end of the light period will give

Silica helps plants resist pathogens including *B. cinera*, the cause of gray mold, by strengthening cell walls making them less vulnerable to fungi. Pro-TeKt from Dyna-Gro is an effective source of silica and potassium.

any excess moisture on the leaves an opportunity to evaporate.

Gray mold does not require the host to be living; it can also thrive on wet and rotting plant refuse. Prompt removal of plant debris reduces potential host sites.

Cut out and remove the affected portions of the plant or bud, or remove it entirely. Destroy any infected material, and sanitize any tools to prevent them from spreading infections. Moldy weed is ruined, and should be thrown out, although unaffected buds from the same plant may still be consumed.

After removing all signs of infection, if at all possible correct the environmental conditions that allowed the colony to start.

If circumstances do not allow the environment to be controlled, preventative measures such as *Bacillus subtilis* (sold under the trade name Serenade), or a fungicide like Ed Rosenthal's Zero Tolerance Herbal Fungicide may be used. Fungicides are more effective when used as a preventative treatment than after a colony has become established, so if used it should be applied at the beginning of the wet season in an outdoor garden. Some gardeners use sulfur to treat at-risk or infected plants. It can be applied as a powder, in sulfur dioxide mats, or from sulfur pots. Garlic, cinnamon, seaweed, and chamomile sprays have antifungal properties and may help control an outbreak.

Gray Mold
(*Botrytis cinerea*)

Look in Part 2: Controls for more detailed explanations of the solutions listed below. Part 2 is divided into four sections, each representing different strategies.

Preferred methods are marked with an ✱ asterisk.

Preventive Measures

This mold is primarily an environmental disease because it is likely to break out under moist conditions, with humidity over 50 percent. Indoors and in greenhouses where humidity can be adjusted, lower humidity using air conditioning, dehumidifiers, or ventilation if appropriate.

Increase air circulation using fans and pruning crowded foliage. Tie branches to open the plant, and widen space between plants.

Barriers & Physical Controls see *Controls: Section 1*

Infected plant removal

Fungicides see *Controls: Section 2*

Baking soda

Bordeaux mixture

Chamomile tea

✱ Cinnamon

Compost tea

✱ Copper

✱ D-limonene

Garlic

✱ Herbal oils

✱ Hydrogen peroxide

✱ Milk

Neem oil

✱ pH up

✱ Potassium bicarbonate

Seaweed powder or liquid

Sesame oil

Silica

Sulfur

✱ UVC light

Biological Controls see *Controls: Section 3*

Bacteria:

✱ *Bacillus subtilis*

✱ *Streptomyces griseoviridis*

✱ *Streptomyces lydicus*

Fungus:

✱ Mycorrhizae

✱ *Pythium oligandrum*

✱ *Trichoderma* species, including *T. harzianum*,

✱ *T. (Gliocladium) virens*

Outdoor Strategies see *Controls: Section 4*

Brassica residues

Leafhoppers

At a glance

Symptoms

Tiny hard white dots protruding from the underside of leaf tissue; small light green leaping insects that jump from plants; sticky residue of honeydew; sooty mold growing on honeydew; wilted or depleted plants; yellow or brown spotting on leaf tips or leaves that curl and fall off prematurely; stippling that causes a white pattern on the upper sides of leaves.

Actual size

1/6" to 3/16" • 2 to 5 mm

Leafhopper getting ready to jump.

L eafhoppers are tiny, wedge-shaped insects that make their living sucking plant juices. They are members of the order Hemiptera and are therefore considered true bugs. Aphids, whiteflies, and mealybugs are in the same order. They belong to the family Cicadellidae, one of the largest families of plant-feeding insects. They number more than 20,000. That's more than all the mammals, reptiles, amphibians, and birds combined!

They are named for their ability to hop great distances. They spring from disturbed foliage in a panicked flurry and either hop or fly away to safer territory, startling the discoverer. They also infrequently run sideways when startled.

Leafhopper adults hold their wings over their backs in a steeple shape, making them appear taller than wide. Enormous bulging compound eyes are positioned on either side of their head, and powerful back legs wait to propel them into the air when danger approaches.

Rows of spines run along the backs of their powerful back legs in patterns that vary by species, providing entomologists a key characteristic for differentiating between them. Leafhoppers are positively identified under higher magnification by their mouthparts or by their genitalia.

Leafhoppers have two sets of wings. The front wings are membranous at the tips, thicker at the base and, when closed, provide protection for the thinner, more fragile rear wings. The young, or nymphs, are very similar to the adults except they have no wings only "buds" at the site where wings will one day develop.

Found on every continent, except Antarctica, and in nearly every environment, leafhoppers vary in length from ⅛ inch to nearly ½ inch, depending on the species. Some are very colorful ranging from aqua to deep green to drab gray. Many have vivid patterns that appear as stripes, waves, or etching-like markings on their heads or wings.

LEAFHOPPER BASICS
Jumping

Many gardeners have been startled by tiny insects propelling their bodies into the air with powerful jumps that take them a foot or more away in a diagonal upward direction. The leafhoppers' jumping is the result of an ingeniously designed system of sensors, joints, and muscles in their hind legs, which are twice as long as the other two pair of legs.

When a leafhopper, peacefully feeding on foliage, detects a change in its environment that signals danger, it crouches down, bending its legs and storing up an incredible amount of energy in its joints. Once these muscles are fully contracted, hairs along the insect's leg-sections touch. The message that danger is imminent is sent from sensors in the insect's central nervous system. The brain sends a message to the central nervous system that is forwarded to the muscles: jump! The leafhopper launches away from the feeding spot at a top velocity of nearly 10 feet per second (6.8 miles an hour) in less than three thousandths of a second.

Feeding

Both adult and nymph leafhoppers have piercing-sucking mouthparts, collectively called rostrums, consisting of two pairs of tubes. The first pair has sharp teeth for piercing plants, and the second pair, the stylets, are sharp-pointed tube-like structures with which leafhoppers suck out juices. Once the stylets are firmly plunged into plant tissue, some leafhoppers secrete a shield of saliva that seals their stylets into the plant cell while they siphon out the goodies. Others pierce cells, excrete watery saliva to begin the digestion of plant tissues, and then move their stylets around, slashing open cells and sucking out the slurry of half-digested contents.

Leafhopper damage is usually seen as "hopperburn," yellow or brown spotting on leaf tips or leaves which causes the leaves to curl and fall off prematurely. It also causes stippling, a whitish pattern on the upper sides of leaves. The exact type of damage varies by species. An infestation of leafhoppers causes wilting, stunting, and mottling of plants.

Developing nymphs cast off their exoskeletons, which can then be seen on the undersides of leaves as tiny white dots clinging to the leaf tissue—evidence that leafhoppers have arrived.

Leafhoppers are greedy eaters and suck copious amounts of plant sap from their hosts. Plant sap is very dilute, so leafhoppers ingest large amounts. Their bodies filter the nutrients and the bugs excrete the rest as honeydew, a sugar-rich liquid. Leafhoppers are not the only insects that produce honeydew. Aphids and other Hemiptera also produce the thick sweet substance. Leafhoppers, however, have added a unique twist to the excretion process. They produce a bubble of honeydew from their posterior end and when this bubble reaches the right size, it is released from the body with an audible "pop," providing leafhoppers with yet another colorful name: sharpshooters.

Large amounts of honeydew causes problems. It drops onto foliage, greenhouse benches, walkways and mulch, creating a sticky mess and a pest-welcoming environment. Other interested parties soon arrive. Ants collect and feed on honeydew. They are sure to find this abundance of food and decide it is the perfect nesting site. Once they take up residence, they may decide to bring along some friends, namely their own herd of honeydew-producing aphids or mealybugs—visitors you do not want.

Sooty mold is a collective name for several fungi that grow on honeydew. The type of fungi forming growth of sooty mold depends on the environment, the type of insect that produced the honeydew, and the host. As the molds grow on the nutrient-rich honeydew, the dark mycelia of the fungi give the surface the appearance of being covered with soot. Sooty molds do not infect plants, but a substantial growth of mold on the leaves blocks light from the leaves, stifling photosynthesis, and causing the plants to weaken and die.

Aside from the mechanical damage leafhoppers do to plant tissue, they are also vectors of viral and other fungal or microbial diseases.

Sex Life

Leafhopper males are quite the romantics. They advertise that they are available by singing. Their songs vibrate along plant stems instead of through the air and only female leafhoppers can hear, or rather feel, their performance. The songs are too low-pitched and soft for human ears, but the female leafhopper has no doubt as to the vocalist's intent.

Once he finds the perfect gal, the male leafhopper mates with her. Female leafhoppers can monitor the sperm and detect if the right amount to fertilize the eggs is present. If not, she will allow him to mate with her again. Some leaf-hopper females remain monogamous and mate several more times with the same male, a rare behavior among insects.

Life Cycle

When the female leafhopper is ready to lay her eggs, she chooses an appropriate leaf vein, the thinner the better, makes a tiny slit with her ovipositor, and lays the eggs inside the vein. These eggs either hatch within a few weeks or remain dormant over the winter, depending on the species of leafhopper and environmental conditions. In such places as greenhouses, indoor gardens, and in parts of the country that stay warm year-long with no dormant period, multiple generations are produced throughout the year.

Once hatched, the young nymphs crawl out of the leaf vein and seek out a new vein, sink their tiny stylets into the tissue, and begin to feed on the cell contents. During the next few weeks, they undergo several molts, shed their exoskeleton—the outer thick skeleton made from chitin—and grow a new, roomier one. This type of life cycle is known as incomplete metamorphosis because the young look very much like the adults. This is in contrast to other insects such as butterflies where the young pass through stages that look nothing like the adult. After the last molt, at which time the wings become functional, the nymph is an adult that hops and flies, seeking out new food sources and mates.

Water retention is a good idea if you're an insect with an exoskeleton. Leafhoppers have their own unique way of avoiding desiccation. Adult leafhoppers coat their bodies with a light layer of a waxy, water-repellant substance that dries to a white powdery appearance. It keeps excess water out of their bodies, preserving the delicate balance of fluids within. In some species, the female stores these brochosomes on her wings, then powders her egg clutch with them after the eggs are laid to protect them from waterlogging, desiccation, and predators. Many insects produce materials for coating themselves, but the egg coating is limited to some species of leafhoppers.

SPECIFIC LEAFHOPPERS

The following leafhoppers are known to affect marijuana and hemp plants.

Glasshouse Leafhopper (*Zygina pallidifrons*)

A persistent pest in greenhouses or any contained growing area, the glasshouse leafhopper is the bane of indoor growers. Adults are about ⅛ inch long. Their bodies are a pale yellow with gray patterned markings on their backs.

Potato Leafhopper (*Empoasca fabae*)

About ⅛ inch in length, the potato leafhopper is green with small yellowish, pale, or dark-green spots along its length. Its host range is extensive: more than 100 plants. These leafhoppers are travelers and spend the cold months in the Gulf states where they feed on persistent plants there, then move north as temperatures increase.

Red-banded Leafhopper (*Graphocephala coccinea*)

This leafhopper is striking in appearance, because of its coloring and its size. Although nearly ⅓ inch in length, the red-banded leafhopper is slender compared to some other leafhoppers, with a yellow head and wings colored with alternating bands of magenta and greenish-blue with yellow margins. Nymphs' color ranges from yellow to green. Red-banded leafhoppers are found from Canada south to Panama. They overwinter as adults in leaf trash on the ground. The females

Potato leafhopper (*Empoasca fabae*).

Redbanded Leafhopper (*Graphocephala coccinea*).

lay eggs in plant tissue in the early spring. While their feeding technique is similar to other leafhoppers, their damage is rarely serious. However, they are vectors of several plant viruses.

"Flavescent" Leafhopper (*Empoasca flavescens F.*)

This leafhopper is known to infest European hemp and is also found in the United States. It closely resembles the potato leafhopper but its color is much paler and nearly white.

Rose Leafhoppers (*Edwardsiana rosae L.*)

Rose leafhoppers are about ⅛ inch in length with a broader head than some leafhoppers and wings that are whitish-green. They overwinter as eggs, tucked safely away under the bark of roses, including the widely distributed multiflora rose. Outdoors, rose leafhoppers are adequately controlled by their natural enemies—assassin bugs, ground beetles, predator mites, and spiders. Indoors, their presence quickly turns into an infestation. With warm temperatures, abundant food, and no predators, they produce generation after generation, each exhibiting logarithmic population growth.

Versute Sharpshooter (*Graphocephala versuta*)

The versute sharpshooter is about ⅛-inch long. Its wings are streaked with greenish-blue and orangish-yellow stripes. On the back of the head are thin, black markings resembling etchings (hence the genus name *Graphocephala*). They have pale blue margins along the outer margins of their heads. Adults of this species overwinter. In the early spring, adults feed primarily on woody plants, but move to herbaceous plants later in the year. Although this species has the same feeding habits as other leafhoppers, it does minimal damage because it doesn't build a large population base. This leafhopper is a frequent visitor to marijuana-research plots at the University of Mississippi.

Agalliota constricta

More nondescript than some other leafhoppers, this species is pale gray or brown, about ⅛ inch in length, and identifiable by two dark spots on its broad head. It is also a frequent visitor to the marijuana test plants at the University of Mississippi.

LEAFHOPPER RELATIVES

The close relatives of leafhoppers that are listed below all have more than a passing interest in cannabis.

Spittlebug (*Philaenus spumarius L.*)

This bug is a close relative to leafhoppers but larger. Spittlebugs are most recognizable in the nymph stage as they may resemble tiny, green frogs. These ¼-inch nymphs station themselves in the crotch of a leaf or branch and surround their bodies with a frothy white material resembling spittle, which hides and protects them from predators as they eat and grow.

Sharing the same incomplete metamorphosis as leafhoppers, spittlebugs hatch from overwintered eggs laid on the undersides of leaves in hardened froth-encrusted batches of two to

30. They move through several stages of development until the final stage when the nymph, no longer hiding in spittle, now resembles the adult except for a dark brown coloration. One generation a year is produced. Spittlebugs are piercing-sucking insects and create mottling on the upper sides of leaves from their feeding. Their damage is rarely serious and the sight of a few of these is not cause for concern.

Buffalo Treehopper
(*Stictocephala bisonia* F.)

Buffalo treehopper adults are green, about ⅜-inch long and humpbacked like the buffalo of the American plains. Evolved to emulate the appearance of leaves, thorns, or bumps on tree branches, treehoppers use this clever form of disguise as their defense from predators. Like their cousin the leafhopper, they too jump and hop when disturbed.

Females have a knife-like ovipositor they use to cut tiny slits in branches and insert up to 12 eggs. It is this ovipositing in the plant tissue that causes the major damage to cannabis plants. The eggs overwinter in their snug cocoon until they hatch in spring as nymphs. Nymphs are brown and resemble their parents without wings. They crawl down from their hatching to feed on grasses and weeds. During this six-week time, they pass through several stages of incomplete metamorphosis, each stage more mature than the last, until they are adults. Then they return to the trees in camouflage.

Both adults and nymphs are piercing-sucking insects that feed on plant sap. Outdoors, they are often controlled by natural enemies such as birds, spiders, toads, mantids, and other predators.

VIRUS TRANSMISSION

Of all damage associated with a leafhopper infestation, by far the most serious consequence of their presence is the transmission of plant viruses. The order Hemiptera, which includes leafhoppers and aphids, contains the largest group of vectors, or insects that transmit viruses. These piercing-sucking insects accumulate and transfer virus particles on their stylets as they hop from plant to plant and feed. Some insects may also accumulate viruses within their bodies as they feed and then transfer these viruses to plants through feeding.

Aphids top the list as having the largest number of vectors in their numbers, but leafhoppers come in a close second. About 90 percent of leafhopper species are vectors.

CONTROLS

Leafhopper controls vary in success depending on leafhopper type. Specific species of leafhoppers respond better to some methods than others. Knowing the leafhopper species based on region and appearance provides information on the best control to use. If a gardener is unable to identify the specific leafhopper type, it may be necessary to rotate through controls until an effective one is identified.

Companion Plants

Companion plants can be used to form a repellant barrier that pests are unwilling to cross. Heavily scented plants such as mints, alliums, petunias, geraniums, and herbs such as oregano, thyme, and rosemary produce resins and essences that repel or kill a wide range of pests, including leafhoppers.

Insecticidal soap such as Safer Brand Insect Killing Soap is a good general insecticide effective against leafhoppers.

Pesticides

Pyrethrum is a safe, effective, and environmentally friendly pesticide that can be used in combination with neem oil or insecticidal soap as a broad-spectrum insecticide that works against leafhoppers and several other common insects. Pyrethrum breaks down readily when exposed to UV light. Insecticidal soaps are effective when applied every seven to 10 days. Insecticidal soaps are safe to use right up to the day before harvest.

Herbal oils from a range of plants can also be used. Cinnamon, citronella, citrus, clove, eucalyptus, lavender, lemongrass, rosemary, and thyme work well as repellants for leafhoppers. A wide variety of herb-based pesticides are available from organic gardening suppliers.

Biological Controls

BUGS

Minute pirate bugs (*Orius insidiosus*) are an option for the potato leafhopper as well as for fungus gnats, thrips, mites, whiteflies, and aphids. These true bugs are shipped as adults, and once released will fly short distances around the indoor garden seeking prey, mating, and laying eggs.

Barriers

Leafhoppers can't feast on what they can't reach. Barrier protections are not practical in all applications, but where they are, it can be effective to place a thin layer of cheesecloth or horticultural Reemay fabric over the plants. In greenhouses or indoor gardens, filter vents to prevent introducing the pests.

Leafhopper Controls

Look in Part 2: Controls for more detailed explanations of the solutions listed below. Part 2 is divided into four sections, each representing different strategies.

Preferred methods are marked with an ✻ asterisk.

Preventive Measures
Insect netting prevents leafhoppers from becoming established in your garden.

Barriers & Physical Controls see *Controls: Section 1*

- Air-intake filters
- Aluminum foil
- ✻ Insect netting
- Sticky cards

Pesticides / Fungicides see *Controls: Section 2*

- Azadirachtin
- Herbal oils
- Hydrogen peroxide
- ✻ Insecticidal soap
- Neem oil
- ✻ Pyrethrum

Biological Controls see *Controls: Section 3*

- Bacteria:
  - Spinosad
- Bugs:
  - ✻ Minute pirate bug *(Orius insidiosus)*
- Fungus:
  - *Beauveria bassiana*

Mealybugs & Scales

At a glance

Symptoms

Whitish, cottony larvae or bugs in the plant nodes; brown or tan-colored wart-like bumps on the stems or along the veins on the undersides of leaves; sticky honeydew residue or the development of sooty mold due to excess honeydew; plant wilt.

Actual size

Scales: 1/10" to 1/2" • 3 to 15 mm

Mealy bugs: 1/10" to 1/5" • 2 to 4 mm

Scales attached to stem.

Mealybugs and scales may not bring an immediate visual to mind in the same way as other pests in this book. However, if you have had much interaction with houseplants, you may have encountered them at some point. Mealybugs are a type of soft-bodied scale. They are like little tufts of cotton, small white roly-poly bugs that on closer inspection have a strong grooved pattern of white or gray waxy threads. Scales are mobile when young, and not different looking than mealybugs, but after maturity, they appear as lifeless brown or tan bumps along plant stems or the veins along the undersides of leaves. You could say the adult scale is the couch potato of the insect kingdom, since it finds an attractive food-rich location and takes up permanent residence, encasing itself so thoroughly that it is often mistaken for a plant disease rather than an insect. Underneath their shell the interior is white and cottony, like the mealybug. They became commonly known as scales due to the resemblance to fish or reptilian scales.

There are many families of scale insects, but they are divided into two groups: armored scales and soft scales. Armored scales produce a waterproof shell that is composed of remains of their old shells, host plant material, and fecal matter. Some species cover it with waxy filament that is layered into a covering. Diaspididae (armored scale) is the family of scales that most often presents problems for cannabis.

Soft scales also produce a waxy covering but it remains connected to their bodies, somewhat like hair. Coccidae (soft scales) is the superfamily name that encompasses the 8,000 species of scales. Mealybugs are a specialized member in the soft scale group and belong to the Pseudococcidae family.

Planococcus citri (Risso): citrus mealybug

Mealybugs cover themselves in a waxy sheath that they exude. They do not form the hard shell that characterizes other types of scales.

Scales are in the same order as aphids and whiteflies, Hemiptera. Mealybugs and scales have very similar lifestyles. Females pick cozy spots on leaves, branches, or crevices. Then they settle down, remain stationary, and suck plant juices. As you can imagine this does not take a lot of intelligence, but these bugs have developed successful survival strategies. They grow hard shells and have large families.

There are many species of scales but the ones most likely to attack your plants indoors are the hemispherical (*Saissetia coffeae*) and the brown (*Coccus hesperidum*), which are both soft scales. Other scales, often with a cottony covering, also occasionally attack cannabis. The brown and hemispherical scales use camouflage to evade detection. Until you notice them as distinct organisms, they appear to be little bumps on branches or leaves. Cottony scales produce a waxy cottony substance that covers their surface and is attached to their bodies.

Mealybugs produce a cottony web that some potential predators avoid. They cover themselves with this web of waxy material. Some species have developed a symbiotic relationship with ants similar to that of aphids. Ants protect and herd them to collect the honeydew.

Three species of mealybug are likely to infect indoor gardens and greenhouses. They are the citrus mealybug (*Planococcus citri*), the long-tail mealybug (*Pseudococcus longispinus*), and the obscure mealybug (*Pseudococcus obscurus*). For the most part, the same controls can be used for all the species. In addition, parasitic wasps that are specific to various species are available.

All members of the order are true bugs. That is, they feed by sucking plant juices through a specialized mouth called a proboscis that is

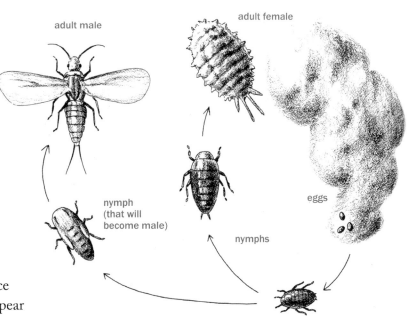

adult male

adult female

nymph (that will become male)

eggs

nymphs

shaped like a straw. Mealybugs and scales are not usually a problem in flowering marijuana gardens because they are relatively slow breeders. It takes a month or more for each generation, so if a fecund female travels to a garden and lays eggs, the first generation won't be ready to breed for five or six weeks. Then a week or two after the second generation hatches, the plants are picked. Compare this with mites or aphids, which have new generations every 10 to 15 days, creating an exploding population very quickly.

Like aphids and whiteflies, scales and mealybugs process and condense a lot of plant juice. They exude this honeydew, which is prized by ants. If there are no ants to eat it, the honeydew is quickly colonized by sooty mold. The plant is weakened both by the insects' leach-like action on vital juices and the herbivore droppings that promote mold infections on the stems and leaves.

The amount of honeydew produced from a serious infestation is amazing. The Biblical manna, which miraculously appeared each morning to be collected as

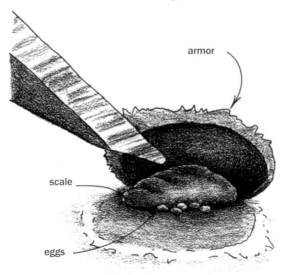

food that sustained the wanderers, was honeydew from a mealybug, *Trabutina mannipara*, endogenous to the Middle East, according to some accounts.

The eggs of the mealybug *P. longispinus* are laid under its body. The female then dies and its body provides a protective cover for the developing eggs. In a week to 10 days "crawlers" emerge from under their dead mother. For this reason a myth developed that mealybugs produce live births rather than lay eggs. Other mealybugs lay their eggs into a cottony sac that they deposit on the plant or carry with them until the eggs hatch.

The crawlers look like miniature females 1/10th the size of an adult. They move over the plant until they find a suitable place to insert their proboscis, then start sucking. Although they look the same when they hatch, males and females are dimorphic, i.e. they have different forms and life histories.

Female mealybugs have three instars or molts, then emerge as adults without much change in form. They search for a suitable spot to suck plant juices and settle down. Members of some species lose their legs and become immobile. When impregnated, they lay eggs under their body or in cotton-covered masses, then die. The mother's dead body becomes the eggs' shelter against predators. The number of eggs varies widely both by species and by temperature and food. Anywhere from 20 to 200 eggs are laid. Their entire life cycle, with good food and warmth, is 25 to 35 days. In cool weather, eating low-nitrogen food, their life cycle takes considerably longer to complete. They are usually not the crop-threatening menace of faster breeding arthropods.

While females have an incomplete metamorphosis, even retaining some of their nymphal characteristics, mealybug males go through a complete metamorphosis. They have two nymph stages over a period of two to four weeks, depending on the quality of the food and the temperature. A high-nitrogen diet found in plants growing in well-fertilized marijuana gardens, and warm temperatures in the 70s and 80s °F, promotes fast growth.

Then they go into a pupal-like stage. They emerge looking something like a mosquito. The male mealybugs are mobile with a single set of wings that are adapted for short flights, ideal for approaching suitable females to impregnate before dying. The back set of wings devolved into stumps. The males are born without mouths so they have physical anorexia and cannot verbalize—that leaves only sex. During their 36-hour adulthood, their sole interest is getting it on. Being mobile, they can search for willing females hunkered down in the branches.

Scales life cycles are similar to those of mealybugs. However, many types of scales, including hemispherical scales, are parthenogenic—which means they reproduce without males and all of the young are females. A single female lays up to 1,000 eggs, which go through several instars before winding up as adults ready to reproduce in 40 to 60 days under greenhouse and grow room conditions.

The brown scales hold eggs inside their bodies during most of the incubation period. They are released shortly before hatching. The female molts twice before being able to reproduce in about 40 to 50 days. Cottony scales have life cycles that span 60 to 90 days. They lay their eggs in masses that they deposit on the plants or, in some cases, remain attached to their body.

Both scales and mealybugs can be a problem in the mother room, which should be kept 100 percent sterile. Here, even one lonely scale or

Diaspididae (armored scale).

Hemispherical scale.

mealybug is too many. Because of their breeding cycle, they are more a threat here because their numbers build up over a period of months rather than weeks. They should be eliminated, not tolerated as part of life as many indoor garden books suggest.

Mealybugs are sometimes found on the underside of leaves, but more often they plunk themselves down into the protected crevices of plants where they are harder to detect by predators and they don't have to worry about the backfield. Cannabis doesn't offer the protected spaces found in some other plants, so the bugs plunk themselves at the nodes where leaves join the branches or branches shoot off the stem.

Scales are found on leaf surfaces, stems, and in crevices. Occasionally scales or mealybugs colonize the stem right at the soil level. Most sources advise indoor gardeners to control mealybugs and scales. Some gardeners are reconciled to living with the pests because they are hard to reach, slow to spread, and their damage is minimal if the population is relatively small. Houseplant gardeners often consider them a nuisance and use physical methods of removal, such as a force-ful water jet, removal with a sponge or cloth, or Q-Tips dipped in alcohol.

I had a bad case of mealybugs on some orchids, which are perennials. I would wash and swab them away, and I used various pesticides including pyrethrum. The mealybugs always returned. However, I finally got rid of them using Ed Rosenthal's Zero Tolerance three times, once every three weeks.

Since a mealybug or scale infestation won't get very far in 60 to 80 days, cannabis gardeners usually aren't too concerned about eliminating them. Eliminating most mealybugs seen in the first 30 days of flowering stops a population explosion that can get serious towards the end of flowering.

Mealybugs and scales in the mother room are another story. A persistent infestation there has the potential of becoming a serious problem because the population explosion occurs over a longer period of time. Mealybugs have the time to reproduce several times on mother plants, which have a much longer life expectancy than flowering plants. The results are weakened plants, the chance of spreading viral plant diseases, and infected cuttings. This is not a pretty sight.

Pseudococcus viburni: obscure mealybug

Pseudococcus longispinus: longtailed mealybug.

CONTROLS

Mealybugs are relatively easy to eliminate on marijuana plants because the plant's structure does not offer easy places for them to hide and protect themselves. Eliminating mealybugs from your garden may be as simple as spraying them off with water, or sponging plants off with a cloth or sponge if the garden is fairly small. Q-Tips can be dipped into rubbing alcohol and then used to clean plant crevices where mealybugs might try to hide. A more serious infestation may warrant the use of alcohol or herbal sprays. Plants can also be misted with oils. Neem and Ed Rosenthal's Zero Tolerance are particularly effective against mealybugs and scales. Pyrethrum may also be used.

Biological controls can also be used, and these strategies may be especially useful for the mother room, where ongoing mealybug and scale control is more important. Some beetles, such as the ladybug *Cryptolaemus montouzieri*, like nothing more than to munch on mealybug larvae and adults. The lady beetle, *Ryzobius lophanthae,* is also a voracious soft scale predator. These beneficial bugs eat some types of armored scales, even though it is not their favorite meal, and they may be especially good to employ in the mother room. However, if the problem is more armored scales than mealybugs or other soft scales, the tiny predatory wasp *Metaphycus helvolus* is the better way to go because it prefers a diet of armored scales. It is an effective control for hemispherical scales in the grow room or greenhouse, but is not as thorough with brown scales.

SUMMARY —
Mealybug & Scale Controls

Look in Part 2: Controls for more detailed explanations of the solutions listed below. Part 2 is divided into four sections, each representing different strategies.

Preferred methods are marked with an ✱ asterisk.

Preventive Measures

Spray with a herbal oil monthly, or every two weeks if pests are present. Use Q-Tips dipped in alcohol to remove bugs.

Barriers & Physical Controls see *Controls: Section 1*

Pest removal using alcohol and Q-Tips or sponge
Water spray

Pesticides see *Controls: Section 2*

Azadirachtin
✱ Herbal oils:
    cinnamon and clove
Horticultural oil
Insecticidal soap
Neem oil
✱ Pyrethrum
Sesame oil

Biological Controls see *Controls: Section 3*

Beetles:
✱   *Cryptolaemus montouzieri*
✱   lady beetles: *Ryzobius lophanthae*
Predatory wasps:
✱   *Metaphycus helvolus*

Powdery Mildew

At a glance

Symptoms
Plants look dusted with powder, often starting with an irregular circle on some small portion of leaves; spreading, it looks webbed.

Powdery mildew spreading throughout the plant.

Powdery mildew is the bane of modern marijuana growers. It is caused by several fungi. It looks like confectioners' sugar dusting leaves. At first it appears as an irregular circle on just a small portion of the leaf. But it quickly spreads onto the surrounding tissue, covering the entire leaf. Colonies soon develop on the surrounding vegetation and in other areas of the garden. Powdery mildew is most likely to attack young leaves, up to two or three weeks old.

The plant becomes infected when a spore, called a conidium, lands on a leaf and germinates. It soon grows an appressorium, a swollen structure that forms at the tip of a strand of hypha that attaches it tightly to the leaf surface. Using the appressorium as a guide tube, the fungus pierces the plant's cell wall and membrane and inserts a haustorium, a projection from the hyphae used to suck nutrients from inside the plant cell. The haustorium sucks up plant nutrients and sends them to the fungus, weakening the leaf and slowing growth. Within a week the fungus produces tiny mushroom stalks that release millions of conidia, ready to infect more leaf surfaces. The fungus also produces another kind of structure, a cleistothecium, which contains spores. It overwinters outdoors and may also hide in a greenhouse or grow room after the crop has been harvested.

THE FUNGI

There are at least two different fungi that cause powdery mildew in cannabis. One of them, *Leveillula taurica,* is probably encountered occasionally in humid gardens. *Sphaerotheca macularis* is the more virulent organism. Powdery mildew species are often specific to a family or species of plants. Hops, closely related to cannabis, is also a host for *S. macularis*.

S. macularis made its debut appearance in gardens throughout Oregon and Washington in 1996. That was the same year that the fungus emerged to attack hops in Washington and Oregon, devastating crops in both states. Until that time, a quarantine had effectively stopped the fungus from migrating to the west from the East Coast and Europe, where it is a serious hops disease. Curiously, the fungus has not jumped from hops to hemp in England and Europe.

L. taurica is more likely to attack warm gardens. It prefers a temperature of about 77 °F (25 °C) and can germinate in low humidity, but functions well in the 40 to 60 percent relative humidity usually found in indoor marijuana gardens. It is inhibited by moisture such as water spray, which destroys its conidia.

S. macularis prefers a cooler temperature. The hops host race grows fastest at 60 °F (15 °C) and stops growing at about 77 °F (27 °C). The virulent race of *S. macularis* found in both indoor and outdoor marijuana gardens today is more tolerant of heat and bright light than the hops race. Both races thrive in moderate humidity and are not injured by water. Their conidia can live in water for short periods and are mobile in it.

In the unheated greenhouse and outdoors, the temperature preferences of both *L. taurica* and *S. macularis* overlap, so plants are susceptible from around 60 to 85 °F (15 to 30 °C). They do quite well in moderate, rather than especially humid, weather.

Conidia of both species are airborne. When

First a few white spots appear.

Spore-bearing organs look like little dark dots on top of the mold.

Filamentary strands of hyphae, close-up.

As the infection progresses, it covers more of the plant.

A thick white powder covers the leaves.

Infected plants lose vitality.

they land on suitable vegetation, they germinate, and start robbing the plant's nutrients. The *S. macularis* conidia also migrate with moving water such as drops of water falling from leaf to leaf, or blown by the wind to other plants.

CONTROLS

Powdery mildew is a plant disorder that is all too familiar to most cannabis gardeners. Because of this pesky fungus' mode of transmission, it can be hard to prevent an outbreak, and is challenging to control once an infection begins. It is in some ways more noxious than insect pests because it ruins any plant matter it touches, rendering it useless. However, infected buds can be processed for their contents to make concentrates such as water hash.

It is important to follow the tips in Chapter 1 on preventive maintenance. Indoors, if you are in an area that is known to have powdery mildew outbreaks, it is especially important to quarantine new plants before introducing them to the garden. Air filtration helps prevent powdery mildew from entering the room in the airstream. If there is any uncertainty about a garden or greenhouse's history with powdery mildew, you should sterilize the area and appliances before starting a new crop.

An additional precaution for powdery mildew is to add UVC light to the garden. With ventilation, all incoming air should pass through the light. UVC light is germicidal, and delivers a fatal blow to all airborne fungal spores that pass under it.

Once an indoor garden shows signs of powdery mildew infestation, there are many different approaches to eradicate it. If it is caught early, the gardener can try simply pruning away any affected plant parts. It is important to disinfect any tools used in the removal of infected plants and to carefully handle the pruned material in order to avoid accidently spreading spores around the garden.

Powdery mildew can also be controlled through close attention to the garden's temperature. Both types of fungi that cause powdery mildew are sensitive to heat and stop growing when temperatures range over 90 °F (32 °C). They quickly perish when the temperature rises to 100 °F (38 °C). Introducing a temperature spike to kill powdery mildew can be implemented, but if it is done improperly it could be detrimental to the plants.

Fungi die in alkaline environments. Adjusting the pH of the leaf surface so it is alkaline, with a pH over 8, is one of the simplest solutions for powdery mildew. Adjustors for pH— one commercial example is called pH Up—can be used to adjust the leaf surface environment to be inhospitable for powdery mildew, clearing up the infection.

Silica helps strengthen the stem, serves as an alkaline adjustor, and has natural fungicidal properties. There are many ways to increase silica in the soil or garden medium, such as adding diatomaceous earth, greensand, pyrophyllite clay, and high-silica fertilizers such as Pro-TeKt. See Controls: Section 2 for more details.

Sulfur has also been used to control powdery mildew for centuries. Several forms can be used. There are formulations that are sprayed on leaves, and elemental sulfur can be used in burners to create sulfur dioxide. Sulfur is more effective at preventing the formation of powdery mildew than treating a moderate-to-heavy infection. When used improperly or at too high a concentration, it causes leaf damage to cannabis

Ed Rosenthal's Zero Tolerance Herbal Fungicide uses potassium bicarbonate to raise pH to a level that inhibits spore growth.

plants. Still, it is an effective method of controlling the infection.

Foliar sprays can be bought or made to effectively control powdery mildew. You can make your own using fungicides such as cinnamon oil or tea, copper, garlic, herbal oils, hydrogen peroxide, D-limonene (citrus oil), milk, neem oil, and vinegar. Some gardeners recommend alternating these treatment options for best results.

Bordeaux mixture is a combination of hydrated lime and copper sulfate dissolved in water. Bordeaux mixture should not be used on the buds. Commercial horticultural oil blends are effective and save you the effort of making your own. Another commercial fungicide that is ef-

Dr. DoRight's dislodges and removes the roots of mildew, disrupting the ability to reproduce.

fective against powdery mildew is potassium bicarbonate.

Some biological controls also work against powdery mildew. The bacterium *Bacillus subtilis* strain (strain QST713) is often known by the brand name Serenade. It is safe to humans and animals, but is highly detrimental to fungi. When applied weekly, it puts powdery mildew into remission. Another strain of bacteria, *Bacillus pumilus* (QST 2808 strain), produces a compound that disrupts fungal development. *B. pumilus* is marketed as Sonata. Sonata does not eradicate powdery mildew completely, but it works well in combination with Serenade or other solutions to greatly reduce an infection.

Powdery Mildew Controls

Look in Part 2: Controls for more detailed explanations of the solutions listed below. Part 2 is divided into four sections, each representing different strategies.

Preferred methods are marked with an * asterisk.

Preventive Measures

Reduce humidity and increase airflow. Treat with *Bacillus subtilis*, potassium bicarbonate, milk, pH Up, or herbal oils. In indoor spaces, the spore population can be controlled using germicidal UVC lamps; this lowers the stress on the plants. Using compost tea or biological sprays are also excellent preventatives. Increase air circulation using fans and pruning crowded foliage, tying branches to open the plant, and widening space between plants makes prevention easier.

Barriers & Physical Controls see *Controls: Section 1*

Heat
Removal of infected plants

Fungicides see *Controls: Section 2*

Baking soda (sodium bicarbonate)
Bordeaux mixture
* Cinnamon oil and tea
Copper
D-limonene
Garlic
* Herbal oils
* Hydrogen peroxide
* Milk
Neem oil
* pH Up
* Potassium bicarbonate
Sesame oil
Silica/silica salts
* Sulfur
* UVC light
Vinegar

Biological Controls see *Controls: Section 3*

Bacteria:
* *Bacillus pumilus* (QST 2808 strain)
* *Bacillus subtilis*, (strain QST713)

Rats

At a glance

Symptoms

Signs of damage to plants in form of chew marks, gnawing, or digging around bases of outdoor plants. Look for signs to positively identify rats as the cause: rat footprints or nests, rat droppings, or dead rats delivered by pets.

Black rat (*Rattus rattus*) in a burrowed nest.

Rats are humanity's longstanding invisible companion, content to hide in the shadows and feast on what is cast aside. This parasitic association goes back millennia and has earned rats the classification of "commensals" for their tendency to live with and depend on man for food. Archaeological finds of rat bones among human remains date back to the mid-third and fifth centuries in the United Kingdom, but the commensalist relationship has existed for at least 20,000 years.

Rats are omnivores. They eat grain, plant matter, carrion, and live animals such as chickens, young lambs, and birds, given the opportunity. The Norway rat will eat soap if nothing else is available. This is one reason they've been able to survive so many concerted efforts to eradicate them.

Another reason for their survival is that rats are prodigious breeders. A single female is capable of producing hundreds of offspring in a short amount of time. Rats often nest near humans without being noticed, although they can show up at embarrassing times.

Fortunately, rats are only occasional visitors in the marijuana garden. They are extremely rare to find as pests in indoor gardens, but may pose an outdoor threat. Rats do not actually feed on marijuana but they damage stems and roots by gnawing and digging. All rodents must grind down their teeth, which grow constantly throughout their lives. Gnawing is the solution. Rats enjoy carrying out this task using the woody stalks of marijuana plants, and they enjoy the accomplishment of toppling plants.

Sometimes rats dig up plants while burrowing to investigate a potential nesting site, or search for food sources based on the odors of blood, fishmeal, or compost. They are most likely to be a problem in the marijuana garden when it is next to something they can burrow under, such as a trash pile, piles of earth or compost, or a barn. Vegetable gardens, cornfields, orchards, and wild areas with abundant nuts and berries provide rats a commissary. Campsites are also a draw for rats if food is left behind.

RATS & CIVILIZATION

Rodents first appear in the fossil record at the end of the Paleocene and earliest Eocene in Asia around 54 million years ago. The original rodents descended from rodent-like ancestors called anagalids, which also gave rise to the Lagomorpha, or rabbit group. Squirrels are also rodents, but members of the family Sciuridae. They appeared at the end of the Eocene era as well. The genus *Rattus* arose about 3 to 4 millions years ago. The black rat (*Rattus rattus*) and the Norway rat (*Rattus norvegicus*) parted company about 2 million years ago. Today, there are 51 species in the genus *Rattus*.

Rats have had a huge impact on civilization. They served as transportation for fleas whose bite spread the plague-causing bacteria that caused the Black Death pandemic, arguably the worst public-health disaster in human history.

In the 14th century, this plague killed somewhere between ⅓ and ⅔ of Europe's population, estimated between 75 and 100 million people. Although the major outbreaks were primarily in Europe in the Middle Ages, there have been outbreaks of plague in North America and other parts of the world during the 20th century. Since 2000, nearly all cases of plague outbreak are in Africa.

Black Death is often used interchangeably with the term "bubonic plague," but it actually consisted of three forms of plague. The other two, the pneumonic and septicemic plagues, were spread in the same way, by bacteria-laden bites from fleas that were carried by rodents. Those suffering from pneumonic plague died the quickest. Septicemic plague turned the appendages of its sufferers black, hence the name Black Death.

Domestication

Rats and man have a checkered past when it comes to domestication. In the 18th century, rats were captured as food in times of famine and as participants in rat fighting rings, or "rat-baiting," in which rats were piled in a pit and gamblers bet on how long it would take a rat terrier dog to kill them. These crude sports gave rise to breeding for competition and also led to the first adoption of pet rats.

White rats were the first animal domesticated for use in scientific research. As lab rats, these rodents, who caused so much destruction, have also contributed greatly to human progress. These small mammals have enough in common with people to serve as good proxies for medical research.

RAT TYPES

In North America, two imported species of rats cause the most damage: the Norway or brown rat, and the black, or roof rat.

Neither species is indigenous to North America. Both are thought to have originated in India and then spread throughout Europe. Members of the order Rodentia and the family Muridae, rats share a common ancestry with other members of this family such as mice, hamsters, voles, and gerbils, but only members of the genus *Rattus* are considered "true rats" or Old World rats. The black rat was present in North America by 1600 but the brown rat didn't emigrate until the mid-18th century. In the ensuing centuries, the Norway rat replaced the black rat and now is the dominant rat by far.

Brown Rat *(Rattus norvegicus)*

This most common and most familiar rat in North America is known by several names: brown rat, Norway rat, Hanover rat, sewer rat, wharf rat, and brown Norway rat. It is one of the largest rats in the order Muridae, measuring an average of 14 to 18 inches in length. Males usually weigh about 19 ounces and females weigh around 12 ounces. They are covered in dense brown or dark grey fur with lighter undersides. Brown rats are excellent swimmers and diggers, but poor climbers.

They are found throughout the world, with the exception of Antarctica, the Arctic, some isolated islands in the province of Alberta in Canada, and protected areas in New Zealand and Australia, making them one of the most successful mammals on the planet, second only to humans.

Black Rat *(Rattus rattus)*

Like its cousin the brown rat, the black rat is also known by a myriad of names: ship rat, roof rat, Alexandrine rat, and old English rat. It followed the Greek and Roman armies out of India to the Mediterranean and reached England by 400 BC. It probably arrived in North America with the first colonists from Europe in the 16th century.

The black rat is smaller than the brown rat, reaching adult lengths of 6 to 8 inches. They have long hairless tails that are about 110 percent of

Brown Rat

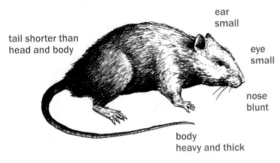

tail shorter than
head and body

ear
small

eye
small

nose
blunt

body
heavy and thick

Black Rat

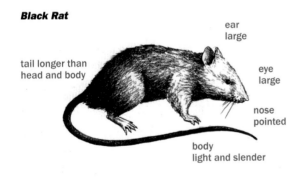

tail longer than
head and body

ear
large

eye
large

nose
pointed

body
light and slender

House Mouse   *Young Rat*

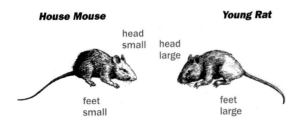

head
small

head
large

feet
small

feet
large

Characteristic	Black Rat	Brown Rat
Color	Predominately black with some individuals brown to grey, underbelly grey to white	Brown upper body with brownish underneath
Size	Slim in build, weighing 5 to 10 ounces	Large, weighing 12 to 19 ounces
Tail	As long as body with fine, black scales	Shorter than body, dark upper and pale underneath with scales
Muzzle	Pointed	Blunt
Ears	Long, will reach eyes if folded over	Short, will not reach eyes
Habitat	Burrows along buildings, beneath rubbish, in woodpiles; indoors remains in basements	Lives and nests above ground in trees, shrubs, and other dense vegetation; indoors remains in attic spaces, walls, false ceilings, and cabinets
Droppings	Larger than mouse droppings, thin	Large, oval, larger than black rat

their total body length—an important characteristic in distinguishing them from the brown rat. Females weigh 5 to 6 ounces. Males weigh about 10 ounces. However, estimates of rat size vary wildly. This inconsistency has given rise to some interesting and amazing folkloric accounts about monster rats. In both Europe and North America, the more prolific and more aggressive brown rat has largely displaced black rats.

Black rats are good climbers, but poor swimmers. In areas where both species of rats are found, general guidelines for identification include where it takes up residence. If the rat is upstairs in a house, it's a black rat. If it's underneath the house, it's a brown rat. Black rats are often seen running along overhead utility lines. They feed shortly after sunset. They gather food and move it to a protected area or to their nest for later consumption. They are particularly cautious of new items in their environment, making trapping them more difficult than trapping brown rats. Black rats share a common biology with the brown rat in nearly all aspects but they tend to produce smaller and fewer litters of pups. The black rat is normally found in warmer areas.

BIOLOGY

Black rats and brown rats share a common biology. Rats are large rodents but rarely weigh much more than a pound, despite urban folklore filled with tales of New York rats that are as big as cats. They nest in dark, quiet places such as burrows, walls, underneath houses, barns, and sewers. They can cause staggering damage. Their incessant chewing and nibbling of walls, floors, and even furniture usually results in much more damage than the loss of the food they consume.

Rats are prolific, polygamous, year-round breeders. Females can mate for the first time when they are 65 to 100 days old. Estrus only lasts about 12 hours and mating usually occurs at night. Intromission is the actual deed and only lasts a few seconds. Once the deed is done, the female moves away and waits for him to mount her several more times over a period as long as 12 hours. The male grooms himself between couplings. When the lovemaking is over, the female grooms herself and the male. The male often makes vocalizations, then becomes sexually inactive and lethargic, lies down, and takes a nap.

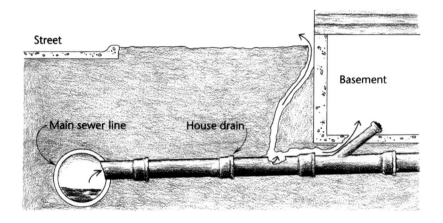

Norway rats can enter buildings through breaks in sewer lines and cracks in concrete foundations.

Street

Basement

Main sewer line

House drain

Rats can't vomit, an evolutionary trait that humans have taken advantage of to kill them with rat poison. Rats have a powerful barrier between their esophagus and their stomach that prevents the regurgitation of stomach contents. In addition to the barrier, they cannot contract the two diaphragm muscles at the same time, which is required to empty the stomach.

To compensate, rats have developed a highly attuned food avoidance sensitivity to prevent poisoning. They take a nibble of a new food. If it makes them sick, they avoid it from then on and communicate the problem to their relatives. If they feel they have ingested toxins or become nauseous, they eat non-food materials such as clay that have the ability to bind up toxins and settle the stomach, passing the toxins through.

THE FOUR RAT SENSES

A rat's world is perceived through four heightened senses, much like ours. Rat vision is quite blurry, around 20/600 by human standards. Rats interpret their immediate surroundings with their whiskers, the long sensitive hairs on their muzzles. They twitch their whiskers back and forth across the air to locate surfaces. This gives them a tactile picture of their surroundings.

Humans and rats both have two types of light receptors in their eyes. One type, called cones, sees bright light and colors. The other, called rods, perceives dim light in black and white, but not color. Humans have trichromatic cones, which blend red, green, and blue to allow us to see the rainbow. Rats, like dogs, have just two types of cones: a short blue ultraviolet and a mid-length green. This is called dichromatic vision. Rats see only two types of light in the spectrum and therefore a much narrower range of color. However, seeing UV is useful because they leave urine traces to mark their territory or advertise their sexual readiness to other rats. The urine is visible in the UV range so they can see the urine as well as smell it. Humans cannot see the UV spectrum but can see urine with the aid of an ultraviolet light.

About 1 percent of a rat's DNA is devoted to their highly developed sense of smell. This is a very high percentage. They must perceive an overwhelming mélange of odors. Every tiny drop of urine left behind by a rat transmits a multitude of information to the sniffer—sex, age, stress level, reproductive status, and sexual availability.

Rats' hearing is sensitive into the ultrasonic range, making it many times mores sensitive than ours and almost as sensitive as a bat's. Humans can hear in the 20 kHz range, but rats hear up to 90 kHz. If you rub your thumb against your forefinger, you hear nothing. But, this motion makes an ultrasonic sound that rats can hear.

In comparison to other animals, here are some hearing ranges:

Humans	20,000 Hz
Dogs	40,000 Hz
Cats	60,000 Hz
Rats	90,000 Hz
Bats	100,000 Hz
Dolphins	150,000 Hz

COMMUNICATION

Rats communicate in both the ultrasonic range that humans cannot hear and in the range audible to humans. Pups communicate with their mother in the ultrasonic range to demand her attention or convey hunger. By 14 days of age, they learn to limit their vocalization in order to avoid attracting the attention of an adult male that may harm them. Adult females also emit ultrasonic vocalizations during mating and when they perceive danger.

Rats have quite a varied vocabulary, although we can't always hear all of it. Chirping is a series of short, sharp bursts of ultrasonic sound. It is the rat equivalent of laughter. Rats chirp in response to receiving something good, such as a belly tickle to a pet rat, or rough-and-tumble play or mating. Rats also make "happy" noises, such as "bruxing," or teeth grinding, in response to pleasurable situations. It is also used for self-comfort during stressful times.

RAT BEHAVIOR

Rats exhibit a fascinating range of behavior, much of it carefully studied and observed by scientists and pet rat owners.

Gnawing

Rat gnawing accounts for much of the damage they do. For these rodents, gnawing is not simply one of life's pleasures, it is a matter of life and death. Since rats must grind down their teeth by gnawing, they are willing to try nibbling on nearly any structure from door frames to electrical wire. Rat teeth grow continuously. If they don't gnaw, the teeth grow too long and they cannot shut their mouths or chew. The situation leads to death from starvation.

Chewing food and gnawing are two different processes to rats. When gnawing, the rat's lower jaw shifts forward until the large incisors are in contact with each other and the back molars are not. This gives them the leverage to nibble plant stalks and fence doors. When chewing, the jaw moves back until the molars are in contact and the lower incisors rest inside the upper without touching. They can then pulverize their food before swallowing.

Digging

Rats dig whenever and wherever they have an option to do so. They pull dirt toward them with their front feet, storing it underneath their stomachs. When dirt has accumulated, they kick it backwards with their back feet. In experiments with domesticated rats, a nest burrow is always constructed at the end of the first tunnel with an escape tunnel leading away from it. In these studies, domesticated rats, bred over many generations for their docile characters, still maintained the ability and instincts to dig safe homes outdoors.

CONTROLS

Rats are amazing survivors and ridding your garden of them can be quite a challenge. If you find you really cannot deter rats through less direct means, you may resort to more straightforward methods. Cats, dogs, and traps are the best ways to eliminate rats. Flooding and smoke cartridges can be used once the nest is located. Baited poisons should be used in controlled areas because of the danger of harming other animals unintentionally.

Four steps are necessary to handle a rat infestation:

- Inspection
- Sanitation
- Exclusion
- Population reduction

How to Spot a Rat

- Droppings around pet food dishes
- Remnants of rat nests underneath firewood stacks
- Burrows among plants or in thick vegetation

Rat in field.

- Burrows beneath the compost pile or beneath the garbage can
- Evidence of digging underneath the garden shed or dog house
- Evidence of rodents feeding on fruit or nuts dropped from trees
- Dogs or cats bringing home dead rat carcasses
- Sounds of scurrying along the sides of buildings
- Tracks in mud

Sanitation

Rats are attracted to places with abundant food and places to hide and nest. Follow the recommendations for outdoor garden preparation in Chapter 1. Your battle to deter rats becomes extraordinarily harder if you inadvertently attract them to the garden where marijuana stalks are susceptible to gnawing. Don't send rats mixed messages.

Traps

Traps are a common method of controlling rats. Several types are available, from the old wooden snap-bar types to wire cage traps. Bait traps with peanut butter, cooked bacon, nutmeat, or a small piece of hot dog. For brown rats, set the traps along observed rat trails, such as along the foundation of a building or known paths in the garden. Set enough traps to do the job quickly as rats will become wary of unsuccessful traps. For black rats, put traps off the ground on ledges, fence railings, shelves, branches, or overhead beams.

Exclusion

It may be next to impossible to exclude rats from an outdoor garden area, but tight fencing, possibly electrified, may be enough to deter rats, because they are really not that motivated to reach cannabis plants as a food source, only for gnawing.

Rats should also be excluded from reaching indoor garden areas. It is easy to:

- Close all openings to buildings
- Close off access along plumbing and electrical lines and stuff holes with steel wool
- Repair any doors or windows that do not close properly
- Eliminate access to compost containers, especially if kitchen waste is composted
- Eliminate nesting locations

Placing fencing around beds or marijuana stalks keeps rats from chewing the plants. Place 1 inch chicken wire, ½ inch hardware cloth, or steel wool around individual plants to a height of 3 feet. This challenges any rat that tries to climb over. This is almost always enough of a deterrent to make them lose interest. Should a rat try heroic actions, line the top of the barrier with a single strand of electric fencing.

Predators

Cats are great biological control for rats and mice. Cats love to roam and hunt outdoors, and their presence makes rats think twice about taking up residence there, creating a good form of biological control. Rodents have a built-in aversion to being near cat odors. When cats spread their scent around an area, mice and rats avoid it.

Dachshunds, huskies, and some terriers are particularly good ratters and can be used to keep the garden free of rats as long as they are in the garden when rats are active. Owls, hawks, and other natural predators lower the rat population a bit, but they will not eradicate an infestation.

Fumigation

Smoke gas cartridges are available commercially to literally smoke rats out of their holes. They contain carbon monoxide and, when ignited, fill the burrows with gas that renders the animal unconscious. Death from asphyxiation soon follows. Use according to label instructions.

RAT CONTROLS

Preventive Measures

A cat or dog can serve as a deterrent so the problem doesn't arise (at least not for long). If an animal isn't the solution for you, use barriers or set traps. Use poisons carefully.

Solutions are specific to rat behavior and how they interact with humans. Good sanitation, exclusion, and population reduction are important. Preferred methods are marked with an * asterisk:

* Animals as pest control:
    birds, cats, dogs. See Controls: Section 4.
  Fencing: rat fence
  Traps

Baited poisons
Flooding
Fumigation

Root Aphids

Root aphids are a serious problem and they are not easy to get rid of. They look like other aphids but colonize and suck juices from stems below the root line or on the roots in many mediums including horticultural pebbles, rockwool, and planting mixes. They often go unnoticed in pest scouting because they live below ground. They are voracious eaters and can do extensive damage if left unchecked. They are most often introduced to growing areas on infected plants or media. Once inside the garden, they begin to suck plants' juices and make more aphids, just like their cousins the "normal" aphids.

Members of the order Homoptera, the most commonly encountered root aphids belong to the families Aphididae and Phylloxeridae. The family Aphididae contains both root aphids and "normal" aphids. The family Phylloxeridae contains the close cousins, the phylloxerans. Aphids from the Phylloxeridae family colonize trees and crops as their primary and secondary hosts, including the infamous grape phylloxera that nearly brought the French wine industry to its knees in the 1850s.

IDENTIFICATION

Root aphids look similar to their aboveground counterparts. They are pear-shaped, light green or white, and about ⅛-inch long. Like other aphids, they have cornicles extending from the rear of their abdomen. However, unlike their cousins, on root aphids these cornicles are ring-shaped instead of the tail pipe structures of the aboveground species, reduced probably to facilitate moving through the soil.

In soil, root aphids are often surrounded by a white, waxy

At a glance

Symptoms

White, waxy webbing that looks like fungus around the roots of plants; wilted, stunted plant, yellowing foliage, premature leaf drop; may be misidentified as a nutrient deficiency because it has symptoms in common and the aphids are underground; if the "fungus" is crawling around, it's aphids.

Actual size

1/8" • 3 mm

Root aphids sucking on root.

webbing that can be mistaken for a symbiotic mycorrhizal fungus that often frequents the roots of potted plants. If the "fungus" is crawling around, it's aphids. Root aphids colonize the roots of potted plants near the edges of the root ball, especially near the drainage holes. They can do extensive damage, especially indoors. Symptoms include wilting, stunting, yellowing, and premature leaf drop.

Root aphid damage is easily mistaken for nutrient deficiencies or other growing-media related problems such as under- or overwatering. As a result many gardens suffer from root aphids but the growers do not know it. There are reports of ants milling around the surface of the soil of root aphid infestations, but I have not seen this phenomenon in marijuana gardens.

The winged form of root aphids is rarely seen, especially in warm, humid growing environments, so there is no aboveground stage to alert the gardener. Adult winged root aphids, or alates, are easily confused with fungus gnat adults when seen. Both are small, winged, and hover among the plants. Fungus gnat adults flit around clumsily. Winged aphids are larger, more sturdily built, and fly with determination, zooming about from place to place.

Biology

Root aphids share a common life cycle with other aphids. Outdoors, females lay eggs on a primary host. The eggs overwinter and hatch into females in the spring. The nymphs molt four times before reaching maturity in eight to 12 days. Then they produce eggs parthenogenically (without sex) that hatch in the reproductive system. They can deliver up to 10 live young a day.

As the population increases in the late spring, hatchlings with wings emerge to become winged migrants. They use the wind to bear them along to another (secondary) host that they feed upon and continue reproducing.

If one of these lucky migrants happens to find a greenhouse or growing room, or even an environment with a milder climate, things get simpler. Now protected from the elements and with abundant food just for the taking, Mama aphids pop out sedentary individuals without wings and no sexual stages—the most commonly seen form.

Root aphids feed in the same manner as other aphids. They pierce the phloem tissue of the plant with their proboscis, which originates between and behind their front legs, and suck out the juices. This proboscis shields two needle-like structures called stylets which are very delicate and damage easily. To compensate for this, the aphid excretes a thin liquid from the ends of the stylets as they begin their insertion. This liquid hardens, forming a protective tube around the stylet as it is slowly inserted. When the stylets reach the phloem tissue, the aphid injects the plant with saliva that prevents the plant from sealing the wound and allows the aphid to feed at leisure. It takes an aphid anywhere from 25 minutes to 24 hours to start withdrawing goodies from the time they first pierce the plant's tissue.

As root aphids live below ground, their presence flies beneath most growers' radar and their damage is mistaken for other problems, or else they are just overlooked. Thus, for now, not much information is available about root aphids and which species feed on marijuana. There are some genera, however, that infest a large variety of plants, and may be the culprits on a cannabis crop.

Pemphigus genera

Root aphids in the *Pemphigus* genera infest a wide variety of crops from lettuce to corn. They overwinter as eggs and infest plants in the spring and fall. They often have primary and secondary hosts for different stages of their life cycle reproduction. *Pemphigus* is a particular pest of greenhouses. Individuals cluster together in colonies and sometimes cover themselves with a white, waxy webbing. They prefer to infest the outer edges of the potted plant's root ball and also prefer the area near the drain holes.

CONTROLS

Before treating any system, whether planting mix or hydroponic, verify the identity of your pest with someone who knows. Root aphids are not fond of damp environments, preferring drier conditions. Therefore, insects infesting roots in a hydroponic system may not be root aphids at all.

Pesticides & Fungicides

Several oils work effectively against root aphids.

- D-limonene: a citrus oil effective against root aphids.

- Neem oil comes from the seeds of the neem tree and is known for its broad-spectrum insecticidal properties. Neem oil is absorbed by the plant roots and works systemically.

- Azadirachtin is an alcohol extract of neem oil that functions systemically.

- Herbal oils are used as soil drenches.

- Apply pyrethrum as a soil drench.

Biocontrols

Because root aphids live nearly their entire lives below ground, especially in grow room or greenhouse settings, standard insect biocontrol organisms, such as lady beetles or lacewings, are ineffective. Only organisms that also dwell in the soil will eliminate an infestation.

These include:

- *Beauveria bassiana*: A fungus that works against a wide variety of pests. It can control over 95 percent of root aphids.

- Beneficial nematodes: Nematodes are microscopic roundworms that parasitize root aphids, thrips, and fungus gnats. Species from the genera *Steinernema* and *Heterorhabditis* are effective against root aphids.

- Spinosad

Root Aphid Controls

Look in Part 2: Controls for more detailed explanations of the solutions listed below. Part 2 is divided into four sections, each representing different strategies.

Preferred methods are marked with an * asterisk.

Preventive Measures

Use benefical nemataodes or *B. bassiana*, which will attack invading root aphids. Apply twice per month.

Pesticides see *Controls: Section 2*

Azadirachtin

D-limonene

Herbal oils

Neem oil

Pyrethrum

Biological Controls see Controls: Section 3

Bacteria:

  *   *Beauveria bassiana*

Nematodes:

  *   *Steinernema* and *Heterorhabditis* species

Spinosad

Root Rot

Six types of root rot plague cannabis. *Rhizoctonia* sore shin and root rot, *Fusarium* root rot, anaerobic bacterial root rot, and Texas root rot are discussed in this chapter. Gray mold and Verticillium wilt are each covered in their own chapters. These diseases are caused by different organisms, but the symptoms of all six have basic similarities: brown rotting roots, leaf chlorosis, shriveled or rotting stems, and wilting—resulting in eventual plant death. These organisms also share common requirements for growth and reproduction: high humidity, cool temperatures, and low air circulation.

Anaerobic conditions severely damage roots, and opportunistic bacteria and fungi cause the roots to rot. By making sure that conditions remain aerobic with well-drained soil, moderate moisture, and oxygenated hydroponic water, these diseases are prevented.

Many of the same controls that work for damping off also help to prevent the development or spread of root rot. It is important to observe sanitation, especially as it relates to the soil—use sterile or pasteurized planting medium, make sure the soil is pH balanced, and avoid planting in cool soils. Outdoors, it is important to avoid areas that have been previously infested with fungal diseases, as the mycelia may persist for up to six years after the last infestation, waiting for an opportunity. Keeping the outdoor garden weeded and free of debris helps reduce outbreaks. Avoid overwatering.

RHIZOCTONIA SOLANI SORE SHIN & ROOT ROT

The brown, shrunken lesions and shredded appearance of

At a glance

Symptoms
Plants wilt, leaves turn yellow, "sore shin," or brown, bruised, or burnt-looking spots; shredding along stems that may turn black; root discoloration. These symptoms foretell the death of the plant in the near term.

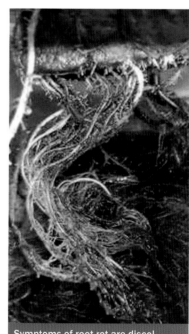

Symptoms of root rot are discolored, soft, mushy roots that often disintegrate.

the infected cannabis stalk gives this disease its unique name. The initial symptom, discolored roots, usually goes unnoticed. Soon after, the compromised roots cause leaf chlorosis (yellowing foliage) and wilting. The disease then moves up the plant stalk above the soil line where several inches of the stem rot away, leaving behind the cellulose skeleton. Within a month, but usually much faster, young plants succumb to the disease and topple over. Older plants may survive but yield is so negatively affected they should be pulled. Small, black sclerotia usually appear in the "shredded" area of the stem.

Rhizoctonia root rot is not limited to cannabis. It is the most economically important pathogen of beans throughout the world and has a broad host range that encompasses most annuals and a large number of perennials. It is particularly damaging to plants that have taproots.

These fungi are everywhere, in all soils, and in non-sterile planting mediums. However, they are opportunistic infections. Roots are much more likely to suffer an attack if they are already in a weakened condition. Conditions that favor infections are:

- Anaerobic conditions. Roots need oxygen to function properly so they are weakened when it is absent, and are susceptible to pathogens.

- Cool temperatures—*Rhizoctonia* thrives in cool soil. It is more likely to be successful when the temperature is lower than 70 °F (21 °C).

Plants won't be attacked if the roots are kept warm and have plenty of oxygen. The easiest way to maintain a high oxygen level is by using a planting mix that drains well. This leaves spaces between the particles that hold air. Containers should be kept warm. If the floor is cool it will drain heat from containers. Place a heat block such as a thin layer of Styrofoam under the containers to keep heat in.

The fungus *Rhizoctonia solani* is the main culprit. It also causes damping off disease of seedlings (see Damping Off). *Rhizoctonia solani* is a basidiomycete. Basidiomycetes are a large group of fungi, and important to humans. The group also includes the rusts and smuts—which account for a large number of agriculturally significant pathogens—as well as edible mushrooms, puffballs, stinkhorns, jelly fungi, and shelf fungi. Basidiomycetes differ from other groups of fungi because their spores, called basidiospores, are produced on the outside of a specialized structure called a basidium, whereas other fungi produce their spores on the inside of their fruiting structures. However, *Rhizoctonia* differs further; it does not produce any asexual spores, called conidia.

Although basidiospores are produced sexually, *Rhizoctonia* still doesn't have much of a sex life. The spore stage of *Rhizoctonia*'s growth is

Roots are rotten and mushy.

classified under an entirely different name: *Thanatephorus cucumeris*, or the "perfect state." *Rhizoctonia* overwinters in soil as sclerotia—dense aggregates of dormant hyphae, the threadlike fungal structures.

Growing plants produce chemicals known as phagostimulants. These chemicals, mostly sugars and complex carbohydrates, attract fungi. In the spring, when air and soil temperatures warm and plants begin to grow, the sclerotia germinate and produce the first hyphae. These hyphae, attracted by the chemical stimulants, grow toward the seedlings. Once they come in contact with a plant, they penetrate the seed coat or epidermis of seedlings by pressuring their way in or entering through wounds.

Once it enters the host plant's tissues, the fungus begins to produce cellulose-degrading enzymes that disrupt the cells of the xylem, the plant system that transports water and nutrients from the roots to the tips of the leaves. *Rhizoctonia* also produces pectolytic enzymes that cause the outside layer of the plant stem to rot away, leaving behind the "sore shin."

Several strains of *Rhizoctonia solani* have been documented on both fiber and high-THC cannabis. The virulence of the disease is not always linked to high humidity and low temperatures. It is also strain-dependent. The disease worsens when conditions are not optimal for the host plant, even when those conditions are also not optimal for the fungus.

CONTROLS

Many of the same controls that work for damping off also help to prevent the development or spread of root rot. It is important to observe sanitation, especially as it relates to the soil—use sterile or pasteurized planting medium, make sure the soil is pH balanced, and avoid planting in cool soils. Outdoors, it is important to avoid areas that have been previously infested with fungal diseases, as the mycelia may persist for up to six years after the last infestation, waiting for an opportunity. Keeping the outdoor garden weeded and free of debris helps reduce outbreaks. Avoid overwatering.

ELISA kits for *R. solani*

You can test plant tissue for the presence of *R. solani* before symptoms are evident. This test, called an ELISA test, contains antibodies which, when in contact with *R. solani*, will turn an indicator color, signaling that the fungus is present. If you are gardening in a new area or one known to have suffered root diseases in the past, this may be a good investment.

Deterrants

Adding high-carbon mulch as top dressing on the soil, or as an ingredient mixed into it at levels up to 20 percent, decreases the viability of *R. solani*.

Adding charcoal at the rate of 5 to 15 percent to the top 6 inches of soil by volume decreases *R. solani* viability.

Rotational planting or early crops of *Brassica* species—broccoli, cabbage, mustard, and wild mustard will, when plowed under, release chemicals into the soil that are toxic to *R. solani*. This effect can be increased by solarization of the soil, a procedure where a soil cover of clear polyethylene is secured over the soil, allowing the soil to retain the sun's heat. The hot temperature improves the release of protectant chemicals into the planting medium.

FUSARIUM FOOT ROT & ROOT ROT

Fusarium solani is the causal agent of *Fusarium* foot rot and, as with *Rhizoctonia* sore shin disease, the first symptoms of an infection are on the roots and usually go unnoticed. As the infection spreads, the stems of plants turn brown at the soil line or the "foot" of the plant.

Aboveground symptoms—wilting and yellowing of leaves—follow. Infested roots turn red, then brown, and rot. This disease is not limited to the young and can kill all forms of cannabis from seedlings to older plants. Do not plant any cannabis crop in infected soil.

Fusarium solani overwinters in crop debris or soil as a resistant spore. Germinating spores infest seedlings or seeds primarily through wounds. Alone, *Fusarium solani* is a mild pathogen, but it is ferocious when it acts in conjunction with plant pathogenic nematodes, such as root knot nematodes, parasitic plants, and broomrape (*Orobanche ramosa*), respectively. (See *Rhizoctonia* Root Rot above.)

Fusarium foot rot has been reported in the former USSR, Poland, southern France, Maryland, Illinois, and Virginia. *F. solani* occurs worldwide. *Fusarium solani* and broomrape together are considered a significant threat to cannabis

The plants dry as the roots collapse.

cultivation. *Fusarium* fungi can pose harm to humans if ingested or if the spores are breathed in. Diseased plants should not be harvested for use.

Fusarium Life Cycle

Fusarium vectors include infected soil, clones, clothing, insects, animals, and even spores floating in the air. *Fusarium* spores are very durable, even in the absence of a host plant. *Fusarium*-infected soil can host viable spores for several years. Spores wait to sprout until suitable environmental conditions occur. After infection, spores spread through water runoff, equipment contamination, plant refuse, and transplanting. Seeds from infected plants can carry the fungus and infect new sprouts at germination.

Fusarium enters the host through the root system, and then spreads upward using the plant's circulatory system. The fungus continues to spread through the plant, causing internal blockages, and restricting water flow. The plant wilts from dehydration. After the plant dies the fungus invades the remaining plant cells, and creates a mass of infectious material. Even if the plant can survive

Remaining roots are white from *Fusarium* hyphae.

to maturity, the buds and seeds are carriers for the fungus, and can further spread the pathogen.

Fusarium Controls

As with other fungal threats, use preventive measures, especially making sure that soil is sterile, in good condition, and well-aerated. In between crops, plow the soil to expose or bury overwintering spores. Keep plants well-pruned and clean any debris. Isolate or remove infected plants and use amendments such as mulch, sand, or compost to improve drainage. In all garden types, avoid overwatering, and avoid creating an infestation through infected seed.

Most importantly avoid infected soils because plants are likely to become infected when they come in contact with this organism. Also, avoid or amend some soils—heavy, wet, or clay soils are especially prone to fungal infections of this type. *Fusarium* infections are also associated with high levels of nitrogen. Balance nitrogen-rich soil by adding potassium and phosphorus to the soil.

ANAEROBIC BACTERIAL ROOT ROT

This disease has somewhat different symptoms, indicators, and solutions than other root rot diseases. Aerobic creatures such as plants and animals use chemical reactions involving oxygen to power metabolism, the life process. Anaerobic bacteria use different chemical reactions so they thrive in an oxygen-free environment.

Plant roots obtain oxygen from their environment, the planting mix, or water. In its absence they become stressed and are more susceptible to attack by pathogens. As a result, an oxygen-

Eliminate Root Knot Nematodes

Root knot nematodes (*Meloidogyne sp.*) are not beneficial but pathogenic. They act synergistically with *R. solani*. Other fungi take advantage of the nematode entry wound to invade the plant. Their elimination from the growing area is essential to controlling *Rhizoctonia* and the other root rot fungi.

Nematodes feed on the roots of plants by piercing the plant tissues with their stylet and sucking out the nutrients. Root knot nematodes set up a feeding site in the roots and get their nutrition from the plant cells around their heads. Once established, the female nematodes become sedentary, never leaving the root. They cause the damage to the root epidermis when penetrating the root to establish a feeding site and this becomes an entry point for *R. solani*, as does the leaky feeding site.

The symptoms of root knot nematodes aboveground are reduced plant vigor, including yellowing and wilting leaves, stunting, and early plant death.

Below ground, the roots are affected dramatically. They may be short, swollen and stubby, and usually sport knots (galls).

In outdoor growing areas, once the nematodes are present in great numbers it is very difficult to get rid of them. Chemical methods don't work well. Crop rotation using plants unfavorable to root knot nematodes such as fescue, onion and other alliums, small grains, Bountiful variety of snap beans, Carolina Wonder variety of bell peppers, and French marigolds can be helpful. Solarization of the soil for several weeks during the summer so that soil temperatures top 110 °F kills root knot pathogens. At 160 °F weed seeds and most fungi are killed. Adding compost, compost tea, and beneficial mycorrhizae help to protect the soil by supplying it with beneficial microorganisms that destroy nematodes.

Probably the best way to deal with this infestation if it is serious is to abandon the soil and instead use raised beds with new planting medium.

deficient environment presents a perfect storm for a successful assault on roots by these bacteria.

Many different species attack roots, and the populations vary by location and the particular environment. Bacteria usually thrive at higher temperatures than molds, from the low 80s °F and up. At higher temperatures, water holds significantly less oxygen than at lower temperatures, so anaerobic bacteria are more likely to thrive in warm water, which is likely to hold less oxygen. If the water is in a planting medium or contains nutrients, it becomes a nutritious soup for bacteria to eat, swim, and reproduce.

You can sense when bacteria are thriving by the telltale astringent odor of the ammonia gas released. When this builds up in the water, it becomes toxic to the roots, killing the plants.

Anaerobic Bacterial Controls

Bacterial root rot is an environmental disease. As soon as environmental conditions change, the problem is eliminated. Problems are caused by overwatering and poor drainage. When the planting mix is filled with water there is no space for air, so the roots do not have access to oxygen. By watering less often, spaces open between the particles. Fresh air fills them. When irrigated too often some planting mixes don't have sufficient time to drain.

Make sure the drainage holes are not blocked and are working properly.

Fill larger containers with coarse-textured planting mix. Then the media won't become compact, squeezing spaces that could hold air.

TEXAS ROOT ROT

This root rot is found only in the southwestern United States and northern Mexico. The causal agent, *Phymatotrichopsis omnivora*, prefers heavy clay soils found in abundance in central Texas, where it has a host range of over 2,000 plants. With little organic matter, high pH, high levels of calcium carbonate, and high temperatures, this area is a perfect habitat. In Texas, Oklahoma, and Arizona, it attacks hemp with 30 percent to 60 percent mortality rates.

The invasion from this fungus begins in the spring with seedlings. Yellowing of the leaves in late summer is quickly followed by wilting, necrosis of the leaves, and plant collapse. Roots necrotize along with plant tops as the fungus invades the xylem and the cortical tissue. Initial symptoms of this disease are often confused with attack by nematodes or soil insects and other fungal diseases.

The fungus grows in rope-like structures called funicles 7 to 10 inches underground. It survives from year to year as sclerotia in the soil that can persist for many years waiting for the right conditions for germination.

Texas Root Rot Controls

Long crop rotation with monocots, such as grasses, corn, onions, and garlic, create unfavorable conditions for *Phymatotrichopsis omnivora* and reduce the incidence of the disease found in the soil. However, the sclerotia of *P. omnivora* persist in the soil for many years, so any crop rotation must also persist for years or the soil in the planting area should be replaced. Another method of thwarting the disease is to use raised beds filled with uninfected planting mix. Avoid both overwatering, and dry conditions. Too little water can predispose plants to infection by stressing the roots. As with other fungal conditions, soil optimization can help. Balance soil nutrients carefully, and, as with *Fusarium*, avoid excess nitrogen by using a balanced fertilizer.

Compost Tea

There are two kinds of compost tea. Both are effective at preventing infections and protecting the plants. Traditionally compost tea was made by "brewing" mature compost in water and then applying it to soil. An enhanced compost tea is made by fermenting microorganisms in aerated water containing nutrients. The organisms multiply quickly and the tea is applied.

Teas are helpful gardening tools because they protect plants from invaders and invigorate plant growth. Directions to make your own are located in Chapter 1: Preventative Maintenance and Controls: Section 2. Ready-made kits can also be purchased, and some gardening stores sell compost tea that is ready for use.

Biological Controls

Many of the same biological controls that work against damping off are also effective against *Rhizoctonia* root rot.

Bacteria: *Bacillus subtilis*
Streptomyces griseoviridis
Fungi: *Trichoderma harzianum*
Trichoderma (Gliocladium) virens
Trichoderma viride

SUMMARY
Root Rot Controls

Look in Part 2: Controls for more detailed explanations of the solutions listed below. Part 2 is divided into four sections, each representing different strategies.

Preferred methods are marked with an * asterisk.

Preventive Measures

Root rot often results from anaerobic conditions, lack of oxygen due to overwatering, or too finely packed planting mix. Water less or replant with an appropriate medium.

Although preventive measures are important for all pests, they are especially important for this particular problem. With fungus, the focus is on sterile, properly balanced soil, and proper watering.

Don't plant in infected soil and prepare soil so that it is well-aerated and drains well. Balance soil nutrients to avoid overly high nitrogen levels.

Eliminate root knot nematodes.

Barriers & Physical Controls see *Controls: Section 1*

Heat
Mulch

Fungicides see *Controls: Section 2*

Brassica residues
(*Rhizoctonia*)
* Compost tea

Biological Controls see *Controls: Section 3*

Bacteria:
* *Bacillus pumilis*
  Bacillus subtilis
* *Streptomyces griseoviridis*
Fungi:
* Mycorrhiaze
* *Trichoderma harzianum*
* *Trichoderma viride*

Outdoor Strategies see *Controls: Section 4*

Brassica residues
* Soil solarization

Snails & Slugs

At a glance

Symptoms

Actual slugs or snails on foliage or at base of plants; slime trails; evidence of chewed leaves, chewed stem bases, snapped plants. This is primarily an outdoor pest issue in zones where slugs or snails live.

Slug feasting on cannabis leaf.

Snails and slugs are two of the most annoying garden and greenhouse pests, especially in humid environments. Not only do they chew holes in the marijuana plant leaves, but they can snap off young plants and, with enough participants, ravage a garden in a single night. They crawl on walkways, greenhouse walls, windows, and other areas leaving behind a silvery trail of dried mucus that betrays their existence and makes affected plant material undesirable.

Snails and slugs are primarily nocturnal, cautious animals that thrive in cool, damp, dark environments such as under leaves, boards, and plant debris. They are not territorial and are content to live in close contact with their neighbors.

Most of the time only about 5 percent of the snail and slug population is aboveground. The other 95 percent is either underground or under leaves, boards, discarded pots, or plant debris, feeding and mating.

Snails and slugs are primarily pests of soft tissue plants, which include the leaves and young growth of marijuana. They vary in size from the miniscule *Punatum pygmaeum*, ⅛ inch long and found in northern Europe, to the whopping *Achatina achatina* of Nigeria and Guinea, with shells more than 11 inches in diameter.

Most pest snails in North America were introduced from Europe, either as food or as hitchhikers.

Cornu aspersa: European brown snail.

PROBLEM SLUGS & SNAILS
Brown Garden Snail (*Cornu aspersa,* formerly *Helix aspersa, Cantareus asperses,* & *Cryptomphalus aspersus*)

This common pest in gardens was originally introduced to San Francisco in the 1850s to satisfy demand for the French culinary delicacy, escargot. They escaped and are now established across much of the United States. These snails have a large, fairly thin shell with chestnut spiral bands and yellow flecks. Adult shells are about 1 inch in diameter.

Brown garden snails mate and then the female lays up to 80 eggs four to six days later. Eggs hatch in about two weeks in warm weather. The young have thin shells and take two years to reach sexual maturity.

European Giant Garden Slug (*Limax maximus*)

Its scientific name literally means "great slug." This huge specimen ranges from 4 to 8 inches long. It is also called the great grey slug or the leopard slug due to black patches that dot its otherwise grey body in leopard fashion. These distinctive patterns of black vary from individual to individual in infinite variability, but its size makes it hard to misidentify.

The European giant garden slug has a complex mating ritual. Two hermaphroditic individuals circle and lick each other for hours. Then they climb to some high location like a tree, suspend themselves from a long thread of mucus and twine their bodies together. Once suspended, they extrude their whitish, translucent penises which then twist around each other and exchange sperm. Sometimes these penises become inextricably entwined. In this event, they chew off each other's penis in a process called apophallation. Now separated, they go their separate way laying hundreds of eggs in dark damp places and continuing life permanently female. If this happens frequently it offers an unusual sexual adaption: All eggs hatch into hermaphrodites. Because individuals lose penises, there is a higher percentage of females to hermaphrodites.

The giant garden slug lives up to three years. It eats dead plant material and fungi as well as living tissue and some of its neighbors. It is listed as a major agricultural pest by many state departments of agriculture.

Grey Garden Slugs (*Deroceras reticulatum*)

The grey garden slug was introduced from Europe during the 1800s and has been spread extensively since then, hitchhiking on ornamental and landscape plants. Colors range from whitish-cream to gray to blue-black. Mottled spots of shades of gray are scattered over it as well. The slime is milky white and sticky. They can travel up to 40 feet in a single night.

As with other slugs, they are night feeders and become active about two hours after sunset. They are considered a serious agricultural pest and lay large groups of eggs. They feed on a variety of vegetable plants, but have a special place in their hearts for strawberries.

D. reticulatum are prey for biological control parasitic nematode *Phasmarhabditis hermaphrodita*, which harbors a mutualistic symbiotic bacterium, *Moraxella osloensi*. The nematode transfers it to the slug when it penetrates its skin. Once inside, the bacteria multiply rapidly and soon kill the slug and digest it. The nematode feeds on the mix of bacteria and dead slug flesh.

Banana Slugs (*Agriolimax sp.*)

Banana slugs are the cuties of the slug world. Some aficionados keep them as pets. Students at the University of California at Santa Cruz chose

Pacific banana slug (*Agriolimax columbianus*)

the banana slug as their mascot. The term banana slug covers three species of the genus *Agriolimax*: *A. californicus* (California banana slug), *A. columbianus* (Pacific banana slug), and *A. dolichophallus* (slender banana slug). They range in color from bright yellow (hence the banana reference) to more mottled hues including green, brown, and white. Some even have spots so dark they appear black. Their bright color is strategic. It says to predators, "I'm brightly colored so I must be poisonous to you!" It's a ruse. Banana slugs are not poisonous.

Banana slugs can reach 9 inches in length, and they can really move, at least in slug terms, inching their way along at 6 ½ inches per minute (33 feet per hour). They are endemic to the northern Pacific coast of the United States and are herbivores, munching on leaf litter, green plants, and mushrooms—their favorite food. They may take a bite out of a marijuana plant if it's available, but it is not their favorite. Nonetheless, you won't have any problems plucking these guys off your plants if they pay your garden a visit.

Other slugs that are common pests in gardens and growing areas include species belonging to the banana slug genus, *Agriolimax*, and "roundback" slugs of the genus *Arion*.

BIOLOGY

Snails and slugs are closely related. Both are mollusks and both are members of the order Gastropoda, from the Greek *gaster* for belly and *pous* for foot. Gastropods are the largest groups of mollusks with more than 40,000 species, comprising more than 78 percent of all mollusks. Snails and slugs are the only mollusks found on land, but they are also found in both fresh and salt water. Most slugs and land snails belong to the informal group Pulmonata, meaning that they utilize lungs instead of gills to draw oxygen from the air.

The lung, or pulmonary cavity, lies on the inside of the roof of the visceral sac. Snails and slugs take oxygen into their bodies through their breathing pore, or the pneumostome. When this pore is open, the top and the bottom of the lung are close together. The bottom drops, increasing the volume in this sac, and oxygen flows into the lung. Then, the pneumostome closes and the bottom goes up again, pushing oxygen into the body.

Snails are comprised of a visceral mass (all of the internal ogans) atop a muscular "foot." The curved external shell encompasses and protects the visceral mass and provides a retreat for the animal when it is threatened. The visceral mass is attached to the shell by a muscle that allows the snail to draw its body into the shell or protrude from it. The coil of most snail shells is clockwise and they are generally carried on the snail's body left-leaning. There are species in which the shell coils in the opposite direction. This trait is often used in identification of species. The shell is composed of calcium carbonate and formed from a thick fold of skin called the mantle. Slugs have the mantle but not the shell.

Both snails and slugs move by sending waves of contractions through the foot muscle. The bottom of the foot muscle is aptly called the sole. Accompanied by the liberal production of slimy mucus, snails and slugs propel themselves safely over nearly any surface, including razor blades or broken glass. The mucus left behind dries to a silvery ribbon. Snail trails are broken while slug trails are continuous.

Most terrestrial snails have four tentacles mounted on their heads that they use to sense their environment. The upper pair of these tentacles (ommatophores) hold their eyes. The lower pair of tentacles serve as olfactory organs. Both sets of tentacles are pulled into the body when the snail feels threatened.

Slugs are snails without shells. They share a common biology. Scientists debate why they are so closely related to snails yet lack a shell. Perhaps they found it advantageous to live underground and the shell that hindered their burrowing disappeared gradually. Perhaps the lack of a shell aided them in crawling into small spaces by being able to manipulate their body.

Snails need high levels of calcium to maintain their shell. Shell-less slugs don't need as much calcium so they can survive in a wider range of habitats. They are, however, more prone to desiccation and must seek out environments that remain moist.

Snails and slugs find food using their highly sensitive sense of smell. They feed by scraping up plant tissue, algae, decaying plant matter, and carrion using their hard tongue called a radula, which contains thousands of backward-pointing, replaceable teeth. Food is smashed against a hard plate in the top of the mouth. Glands near the opening of the mouth release saliva containing digestive enzymes. This predigests the food before it enters the digestive tract.

As both slugs and snails grow, muscles on one side of their bodies develop before the corresponding muscles on the other side. This causes them to undergo torsion, the counter-clockwise turning of the body. It twists the nervous and digestive systems and moves the mantle cavity from the rear to over the head. One theory of its evolutionary advantage is that the mantle positioned over the head helps the snail to draw its head into the shell quicker than if the shell was further back. This does not explain the evolutionary advantage for slugs.

Some snails close the entrance to their shell with a tough structure—an operculum—attached to the base of their foot. The appearance of this structure appeared soon after torsion, suggesting a link between torsion and the development of the operculum.

When land snails crawl, their shell rocks back and forth over their body. Perhaps the anterior positioning of the mantle makes for better ventillation as the shell would block the entrance to the mantle if it were in the rear.

The Slime

Everyone who has ever tried to get snail or slug mucus off their hands understands the unique quality of this substance. The slime is composed of highly organized polymeric material that absorbs water at a rate up to 100 times its initial volume. The slug's cells package the slime as granules encased inside a thin membrane that keeps them dry while the package is inside the cell. Once released by the cell, the packets break open and the polymer quickly absorbs water from surface and the air. The thickness of the slime is governed by the saltiness of the surface.

This mucus serves several purposes. It acts as lubrication allowing the foot to glide over a variety of surfaces. It also protects the snail's delicately balanced bodily moisture from being absorbed by the dry surface over which they crawl.

Slugs and snails are most active in rainy weather and in humid environments. When drought conditions strike, snails use the mucus to protect themselves until the rains return. They bury themselves in the soil or find some other cozy place, such as dead leaves or compost. Snails withdraw into their shells and plug up their shell holes. Slugs secrete a slimy cocoon of mucus that hardens and protects its vulnerable body. Then they slow their metabolisms and descend into a dormant period called estivation to wait until returning rains dissolve the mucus.

Without the protection of a shell, slugs have adapted to use their mucus for defense. When attacked, slugs contract their body making it hard to be eaten. Slug mucus is thicker than a snail's. A predator that snatches up a slug gets a mouthful of slimy, thick mucus. Most of the time the slug is quickly spat out in favor of less offensive prey.

How do you get slug slime off surfaces or, more importantly, your hands? First, wipe the majority of slime off with a paper towel or cloth. Then, wash the remnants away with rubbing alcohol, a mildly abrasive soap, or kitchen cleanser such as Comet.

Snail Sex

All land snails are hermaphrodites. Each snail has both male and female reproductive organs and function as both male and female in sexual encounters. They are not parthenogenic. While they most frequently mate with other snails, they will mate with themselves if no other mates are available. They find mates by using their very acute sense of smell and by following slime trails. In some species, the male and female reproductive organs mature at different rates so a snail may be only a female or only a male for a while, then hermaphrodite later.

Snail sex involves caressing, intimate nipping, and a dance around each other. While the snails flirt and seduce each other, pressure builds up in the blood sinus that surround "love darts," hard, calcareous projectiles coated with mucus.

When snail No. 1 touches the genitals of snail No. 2 the first snail fires his/her dart. Then, snail No. 2 fires his/hers in response. These darts bury themselves in the snail's body like little arrows. Speculation is that this peculiar mating ritual may have given rise to the Cupid's arrow myth. Snails that have been skewered through the head by inaccurate dart shots survive. Many snails, including the brown garden snail, engage in this mating ritual.

Snails digest up to 99.98 percent of the sperm they receive from a mate instead of allowing it to fertilize eggs. Once the dart is embedded in the mate, the mucus on the dart makes its female organs more receptive to being fertilized. The fertile snail digs a hole in the soil with her foot and lays up to 300 eggs, then covers them with a layer of mucus.

Peak times for reproduction are April and May in the spring, and September to November in the fall. In the wild, adult snails seek out damp locations with plenty of food for their impending offspring such as protected garden spots, woods, and pastures.

The eggs hatch in two to four weeks, de-pending on temperature and soil moisture levels. Young snails look just like their parents, only smaller. Their shells are transparent at first, then darken and grow thicker as the young snail eats calcium-containing food such as its cast-off egg casing, or nearby unhatched eggs. The young snail's shell grows as its body does, adding layers to the outside of the coil so that the oldest shell is in the center of the spiral and the youngest on the outside. Snails reach sexual maturity in about two years. Then they seek mates or self-fertilize their own batch of eggs.

Slug Sex

Like snails, slugs are hermaphroditic and fertilize themselves if no mates are around. However, they prefer to trade genetic material with another slug. Their courtship ritual involves similar foreplay, but slugs have forgone the love dart bit and engage in something equally as bizarre— they extrude their penises and wave them over their heads. Other species add another layer of the bizarre and hang upside down by a string of slime and intertwine their reproductive organs, so much so that the entwined organs may have to be chewed off to separate. Others are a little more conservative and just align their bodies so that the reproductive organs match up male to female.

Slugs lay fewer eggs per batch than snails, and bury them in the ground. The eggs are small and round, about 6 millimeters in diameter. Most species lay batches of 20 to 50 eggs. They hatch after several weeks into small slugs that look exactly like their parents. However, eggs laid in late fall may overwinter and hatch when the weather warms. Slugs become sexually mature in less than a year.

Disease Transmission

Although represented as mostly benign organisms aside from their leaf nibbling, snails can transmit diseases both to humans and to other species of snails. The rat lung worm, *Angiostrongylus cantonensis*, which causes eosinophilic meningitis—a potentially lethal human disease—is transmitted by snail mucus on leaves and surfaces. Some species of snails serve as an obligate intermediate host in the life cycle of parasitic flatworms.

THE GOOD GUYS

Despite the furor over introduced species of snails that have become pests, native species of snails and slugs serve an important purpose in the ecology of a garden. Their grazing and digging helps disperse seeds and spores, breaks down decaying plant matter, and helps keep other pests under control. They provide food for opossums, raccoons, birds, turtles, other snails, some beetles, and other animals.

CONTROLS

Slugs and snails like dark, damp places to hide and breed. Eliminate these areas by moving piles of leaves, tree limbs, used pots, stones, low branches and especially, boards. Slugs and snails prefer hiding under flat objects. Thin groundcover to create a drier environment by letting in more light and allowing more air circulation. Eliminate dense, weedy areas around the bases of plant stalks.

Using a cover such as horticultural cloth, or exposing the bare ground deprives them protective cover, making it less likely that they will traverse the area.

Tilling between crops exposes the hidden population, creates drier conditions, eliminates attractive weeds, and encourages slugs and snails to relocate to less disruptive neighborhoods.

Snails and slugs are good candidates for handpicking. They are large enough to see easily, don't bite or sting, and move slowly. The most effective times to search for snails is in the rain or fog, or on cloudy, moist days when they are most likely to come out to feed. Evening is another prime time for snail hunts.

There are two possible problems associated with handpicking. First, it is time consuming and the novelty of snail searches usually wears off quickly, that is, after the first few hunts, leaving only a tedious chore. Secondly, most of the time, most of the population is in hiding or underground.

Snails and slugs are attracted to citrus fruits, so leaving some orange or grapefruit skins in a cool, moist, shaded area draws them to a central location where you can easily collect them.

Leopard slug

Snails and slugs are attracted to yeast, which is why they are drawn to beer. Filling a vertical sided container such as a bottle or can with yeast draws the mollusks. They crawl down the wall and die in the brew. You need not use beer to make a trap. Just fill containers with sugar water and add yeast starter. The yeast digests the sugars, releasing odors that draw the snails. Change the beer or beer starter about twice a week.

Barriers

Snails and slugs are stuck on the ground, although they can scale many surfaces vertically. Moats are effective barriers because they cannot swim and avoid the dip.

Copper is an effective barrier because it interacts chemically with slug and snail slime to produce small, unpleasant electrical shocks that keeps them away. Copper screening or banding for this purpose is available in garden shops. Other barriers include Tanglefoot and Vaseline on cardboard, dry diatomaceous earth, and dry sawdust.

Animal Controls

Domestic animals including chickens, ducks, and turkeys like to snack on snails and slugs. Encouraging wild predators such as foxes, frogs, lizards, opossums, and toads help keep the population under control.

See Controls: Section 4 for additional suggestions.

Pesticides & Fungicides

The best pesticide for snails and slugs is ferric phosphate. The iron content is both attractive and fatal to them. It's a natural product that is safe for pets, humans, and plants.

Biological Controls

The parasitic nematode *Phasmarhabditis hermaphrodita*, marketed as Nemaslug, seeks out slugs and snails in the soil. Decollate snails, *Rumina decollate*, eat snail eggs and immature garden snails, but this biological control solution is limited in its application because it can take a year or more to establish the decollate snail. However, it may be appropriate in areas where snails are a chronic problem and the garden is sizeable and staying in the same location. Make sure they can be used in your region—decollate snails are restricted in some areas due to the threat they post for native populations of endangered mollusks. Decollate snails are illegal to use in California.

Snails & Slugs
Controls

Look in Part 2: Controls for more detailed explanations of the solutions listed below. Part 2 is divided into four sections, each representing different strategies.

Preferred methods are marked with an * asterisk.

Preventive Measures

Keep the area around the garden cleared of debris, and keep grass and groundcover short, so few residential options are available. Place a barrier of ferric phosphate on the ground.

Barriers & Physical Controls see *Controls: Section 1*

    Citrus

* Copper stripping

    Diatomaceous earth: dry

    Moats

* Pest removal: handpicking

    Sawdust

    Yeast: beer

Pesticides see *Controls: Section 2*

* Ferric phosphate

Biological Controls see *Controls: Section 3*

    Nematodes:

*   *Phasmarhabditis hermaphrodita*

    Snail predators:

       Rumina decollate

Outdoor Strategies see *Controls: Section 4*

    Animals as pest control: fowl, foxes, frogs, opposums, and toads

Spider Mites

Spider mites are the most pernicious and persistent pests an indoor gardener is likely to face. The little suckers are not insects, but arachnids, the same class of animals, Arachnida, as spiders and ticks. The reason they are called spider mites is that they string silken webs they use as highways. Two species of mites are most likely to annoy cannabis growers: the two-spotted spider mite (*Tetranychus urticae*), named for the two spots clearly seen on adults' backs at maturity, and the Carmine spider mite or red mite (*Tetranychus cinnabarinus*). They are distinguishable from each other, but are usually grouped together as red mites because both species turn red in the fall as they prepare for diapause (hibernation) in temperate climates. Indoor growers rarely see them turn color, even when light hours are restricted during flowering. The warm temperature in the room keeps them active with no seasonal interruption.

Outdoor growers may see red mites in the fall as the weather cools. They congregate at the tips of leaves and flowering tops in large obvious groups with thousands of individuals clinging to plants and each other in preparation for hibernation.

There are several reasons why gardeners consider mites insidious. First, they are not apparent in the garden until they have already inflicted serious damage. Secondly, they are very small so they are likely to go unnoticed for some time unless you inspect the plants regularly. Third, they multiply so quickly that missing one can spell a new crisis in a few weeks.

Many garden book authors claim that once a garden is mite infested it is impossible to get rid of them. Then they recommend ways to "control" the plant killers. Mite predators can work to keep spider mite numbers to a minimum, but there is no way to really negotiate peace with these critters. They must

At a glance

Symptoms

Early symptoms include "stippling"—small necrotic (browned-out dead) spots on the tops of leaves; parched yellowing leaves that droop, turn brown, and fall off; spider-like webs, egg casings, and little black balls of fecal matter on the undersides of leaves, sometimes starting with the lower leaves and working up.

Actual size

1/5" • .5 mm

Spider mite web covering bud.

The two spots are clearly marked on the back.

Microscopic photo of a two-spotted spider mite.

Close-up of two-spotted spider mites and abundance of eggs on underside of cannabis leaf. Symptoms of spider mite damage (left), which include small necrotic spots on the tops of leaves (right).

be annihilated or they will succeed using the same strategy that insurgents practice: by conducting a war of attrition.

In this scenario, you, the garden, and the plants are "the establishment." In order for you to win—that is, to maintain healthy plants that yield a large crop of healthful desirable buds—you must keep mites, "the insurgents," at the most minimum of levels. Mites, on the other hand, don't have to win to ruin your garden and cause the collapse of your little empire. They just have to not lose. So even if you are "winning" the battle and the mites are making little headway, they still affect the plants negatively. Under the control paradigm, there is constant war and you must be constantly vigilant. The mites have the discipline to eat and reproduce. Do you have the discipline to plan a strategy and to carry it out? It's a toss-up as to who will win.

IDENTIFICATION

To the naked eye, mites appear as black dots on leaf undersides. They can be positively identified using magnification such a magnifying glass or photographer's loupe. Mites are tiny and are more closely related to spiders and ticks than insects. They can be hard to see with the naked eye and harder to detect when the infestation is small. Adult females are about 0.4 to 0.5 millimeters in length and their posterior end is rounded. Males are slightly smaller and the back end of their bodies is more linear than the females. They have eight legs, no antennae, and a single, oval body region like their cousins the ticks. Their feeding structures are two needle-like feeding tubes called stylets with which they pierce plant tissue and suck out the plant juices.

Spider mites are most likely brought into greenhouses and grow rooms inadvertently, transported by gardeners or their pets, or on clones from infested mother plants. Damage from an early infestation is hard to spot. Mites puncture the plants to feed and leave behind tiny, light tan-colored spots on leaves called stippling. The tiny spot is visible from both sides of the leaf. It begins about the size of a pin-prick and then enlarges. Mites sometimes feed in lines parallel to the leaf veins where the juices accumulate.

An infestation quickly reaches epic proportions because of their fast maturation and high birthrate. You know you're in trouble when leaves turn yellow then brown, droop, and fall off. The webs, which are highways between the leaves, are visibly apparent. The mites infest lower leaves first and they work their way up to the tasty, tender growth. Infestations are spotty at first, making the infection hard to detect. Symptoms quickly worsen. Left unchecked, whole plants eventually dry up from the juice sucking. The webs form scaffolding of silvery threads over the entire canopy.

Biology

Spider mites start life as eggs laid on the underside of a leaf by the fecund female. At first they are clear, then turn cloudy, then cloudy-white, and finally, in their last stages, they look shinier and turn straw-colored. In two to four days, depending on temperature, the first larval stage emerges. They look like small versions of the adults except they have only three pairs of legs and are transparent. They immediately puncture holes in the leaves and start feeding on plant juices. Depending on the juices they suck, they turn shades of amber, brown, gray, or green. Mites feeding on

cannabis usually turn amber/gray. Two spots soon develop on the middle of the back of *T. urtica*.

Temperature and quality of diet determine when the spider mite molts, usually within two to four days, taking the form of an eight-legged protonymph. This stage lasts only another two to four days. Then it molts again into a larger version called a deutonymph. A few days later it morphs for the last time. Then it weaves a silken cocoon and pupates for another two days or so.

The time it takes a mite to mature was tested in a series of experiments on maturation and temperature:

Mites Mature Faster in Hotter Temperatures		
Temperature (in °F)	Temperature (in °C)	Days to Maturity (egg to adult)
50–68	10–20	29 days
68	20.3	17 days
81	27	8 days
77 night to 95 day	25 night to 35 day	9 days

While the female mite is maturing into an adult, she produces a sex attractant that leads males to hang around. Fighting occasionally breaks out. This includes bluffing, grappling, pushing, biting, and entangling the opponent in silk. Usually the biggest, toughest mite wins. There aren't many fights, since only a quarter of the mites are male.

Males are produced by unmated females. They only have one set of chromosomes, while females have a pair. As a result, any mutation is immediately expressed in males because there is

Break up the life cycle by spraying Ed Rosenthal's Zero Tolerance Herbal Pesticide three times, three to four days apart.

no mediating second allele. Thus, a favorable new characteristic, such as pesticide resistance, is quickly transferred throughout the population.

When females mature as adults, they concentrate on the two functions necessary for continuing animal life: feeding and reproducing. After a female mates, she lays all-female eggs, but if she finds no lover, she lays male eggs. Depending on temperature, egg-laying commences in as little as half a day, or takes as long as three days after emergence from the pupal state. Maximum egg laying occurs at 77 °F (25 °C), when females lay about 12 eggs a day. Over her 30-day lifetime, she can lay up to 200 eggs, although 100 is a more likely number. At 95 °F (35 °C) the female only lives about 10 days. She lays more eggs daily but lives a shorter life so she produces fewer total eggs. Her aging trajectory slows down considerably as the temperature drops. At 59 °F (15 °C), she lives as long as 40 days, but reproduces at a slower pace.

The incredible number of eggs that female mites lay, coupled with the short time it takes for them to reach sexual maturity, assures a logarithmic population increase. The difference between a small infestation and stressed plants is just a matter of time. A single healthy female can create a plague in a matter of weeks. For this rea-son any mite infestation should be treated as an emergency or a crisis and steps should be taken immediately to kill or remove the suckers.

Famed entomologist John McPartland developed a table of infestation severity for mites. The time it takes a population to move from a light infestation to critical levels is just a matter of a few weeks

Severity Index for Mite Infestations	
Light	Any mites seen—often no symptoms
Moderate	<5 mites/leaf (not leaflet)—feeding patches present
Heavy	>5 mites/leaf—feeding patches coalescing
Critical	>25 mites/leaf—shriveled leaves and webbing present

Credit: *Hemp Diseases and Pests: Management and Biological Control* (2000)

CONTROLS

Spider mites can be eliminated and it doesn't take rocket science to do it. There are various high-tech environmental and chemical weapons as well as biological insect mercenaries that can be brought to the battle against mites. The first step is to assess your risk of mite infestation.

Garden location is the first determining factor. Warm areas surrounded by vegetation are at a higher risk for infection. Cool areas with salt air possess built-in deterrence. I know several gardeners who maintain indoor gardens within eyeshot of the ocean. The salt-air breezes keep pests, especially mites, from the local vegetation so there is less of a chance of infestation.

The second risk factor for indoor gardens is

the points of entry. Since mites don't generate spontaneously, there must be a source of infestation. How were they introduced? Most likely a human or pet carried a mite in or a new plant introduced it to the garden. The ventilation system is another likely culprit. Unfiltered air intake offers no protection against mites. If this source of infestation is continuing, stop it. It doesn't pay to eliminate pests only to have them re-infect. Use a HEPA filter or an insect/mite shield filter over the air intake tubing.

You cannot successfully treat just one section of a garden for mites. The entire garden must be treated at the same time because you will not be able to keep the pests from re-infesting the cleaned part of the garden if there are areas that are still contaminated. This is obvious if the garden is in a single room. However, several grow areas in the same structure or even in separate structures may constitute one garden because of the traffic between them.

Poor sanitation is a related third source of risk. It is important to follow the guidelines for preventive maintenance. Keep plant debris swept up or picked up to eliminate areas where mites live. If possible keep a 10-foot weed-free zone around grow rooms or greenhouses as they are hosts for mites.

Rigorous garden-entry protocols greatly reduce the chance of mite infestation, but they are similar to using birth control—one slip-up and all your efforts can be rendered moot. However, that kind of one-shot luck is rare. Suffice it to say that these policies rely on vigilance to work. Never go from an infected to a non-infected garden without bathing and changing your clothes. If you don't plan on washing your hair place a protective hat or shower cap over it in both gardens.

New plants are not exempt from entry protocols. Dip all new plants in a miticide. Then quarantine them for 10 days until they are cleared of suspicion. These policies are a particularly good

As population increases the mites build webs, which serve as highways, as they move from leaf to leaf. They move in either direction. Does this indicate social interaction?

Red mites congregating.

practice for mites and other plant pests and they improve overall health and pest prevention in the garden. Take care not to introduce mites from mother plants into growing room.

If mites are discovered in the garden, keep their fast-paced life cycle in mind and don't procrastinate. Begin treatments ASAP. Here are some effective methods of treatment to implement.

PHYSICAL REMOVAL

Power washing and vacuuming reduce the population and immediately reduce stress on the plants, but are only a first step in a multi-pronged approach to mite control. These two methods are simple, don't introduce any chemicals or substances to the garden, and bide time before introducing other measures.

PESTICIDES—OFTEN REFERRED TO AS MITICIDES
Herbal Pesticides

Several herbs and spices are effective when used as deterrents.

- Capsaicin, the heat of hot pepper plants, deters spider mites. However the treated leaves cannot be processed to make edibles but can be used for other preparations.

- Herbal oils such as cinnamon, clove, coriander, lemongrass, geranium, rosemary, sesame, garlic, and oregano, or herbal oil mixtures, are extremely effective. You can make your own with a single ingredient or a combination of ingredients. There are a number of commercially produced herbal-based miticides.

- Brewed tea: Rather than using spice oils, you can brew a spice tea by steeping spices or herbs in hot water. Citrus rind contains D-limonene, which is also a very effective miticide. It can be added to the brew. Citrus oil miticides are available commercially.

SNS-217 Mite Control Spray by Sierra Natural Science utilizes rosemary botanical extracts to combat spider mites.

- Some light horticultural oils can be applied as a "summer" spray, but only during the vegetative stage.

- Neem oil, pressed from the seed of the neem tree, is an effective miticide.

Insecticidal soap knocks down populations but is a temporary stopgap because it must be used repeatedly, building up layers on the leaves. It reduces rather than eliminates the problem.

Pyrethrum is only effective against some mite populations. Because of its extensive use in agriculture, most populations have developed immunity to it.

Sulfur burners are effective at eliminating mites. Only some eggs are killed so the burn must be repeated after several days.

BIOLOGICAL CONTROLS

Biological controls can play a key role in eliminating mites. They are effective because they are able to reach all of the suckers.

Bacteria: *Saccharopolyspora spinosa* (Spinosad)

This naturally occurring soil bacterium is useful against ants, spider mites, caterpillars, and leaf miners, but must be ingested to work and is toxic to bees.

Fungus: *Beauveria bassiana*

This fungus also works against aphids and whiteflies, but it cannot be used in conjunction with other beneficials.

Insects & Beneficial Mites

Insects do not eliminate mites from the garden. Under the best circumstances they bring the population down so low that there is effectively no damage to the plants.

BUGS: MINUTE PIRATE BUGS (*ORIUS SPP.*)

These guys love to eat spider mites. Introduce 50 adults for every 100 square feet of space.

MIDGES, GALL: *FELTIELLA ACARISUGA* (CECIDOMYIID GALL MIDGE)

Midges are small fly-like insects. *F. acarisuga* spend their lives seeking spider mites to eat, making them a great choice. This is an excellent insect choice in vegetative gardens. They must be introduced when mite populations are low or reduced.

Many predatory mites specialize in two-spotted mites. While they can be introduced in high concentrations to control a heavy infestation, they are best to use when populations are low. Introduce them after reducing the population using physical treatments or herbal pesticides. The key is matching the predators with your specific environment. Each species has its own temperature, relative humidity, and light needs that must be considered for successful results.

- *Amblyseius californicus* (sometimes identified as *Neoseiulus californicus*).

- *Galendromus occidentalis*: western predatory mite, most tolerant and versatile predator mite.

- *Mesoseiulus longipes:* Although not the most voracious predator, it is tenacious and is more flexible on conditions than many predatory mites.

- *Neoseiulus fallacis*, also known as *Amblyseius fallacis:* good candidate for grow rooms and greenhouses, these mites are tolerant of a wide temperature range 75 to 90 °F (23 to 32 °C).

- Phytoseiid mites *(Phytoseiulus persimilis):* a voracious eater of spider mites, but only works in high humidity environments.

The aroma of certain varieties results in repellant qualities. I asked Subcool of TGA Genetics about comments regarding Jack Cleaner and here is his reply:

"I hesitate to ever make a claim like that but indeed mites seem to prefer strains that have a berry or fruity smell. For instance the Purple Urkle seems to almost spring mites from the plant itself. However the strong lemon astringent smell of Jack Cleaner and its hybrids seem to put the little buggers off somewhat. I have seen gardens fully infested and there were so many less mites on the Jack Cleaner we feel that it must be their least favorite chew."

On a personal note, I have never had much success with beneficial mites. When I have seen them used successfully they were constantly re-introduced. They were not self-sustaining.

Look in Part 2: Controls for more detailed explanations of the solutions listed below. Part 2 is divided into four sections, each representing different strategies.

SUMMARY
Spider Mite Controls

Preferred methods are marked with an ✱ asterisk.

Preventive Measures

One of the most common (and most destructive) pests in cannabis gardens, spider mites can spread throughout a garden at an amazing rate. First, consider your location and the preferences of mites. Does your location allow any options that lower risk of infestations? Second, set rigorous protocols to avoid carrying pests into the garden space. Third, keep the garden sanitary and free of debris. See Preventive Maintenance as a primer or a reminder on important garden basics. Unfortunately, spider mites thrive under the same conditions as the plants, dry and hot. Mist weekly with herbal oils to repel pests.

Barriers & Physical Controls see *Controls: Section 1*

* ✱ Air-intake filters
* ✱ Vacuuming
* Water Spray

Pesticides see *Controls: Section 2*

* ✱ Azadirachtin
* ✱ Cinnamon
* ✱ D-limonene
* Garlic
* ✱ Herbal oils
* Horticultural oils
* Insecticidal soap
* Neem oil
* Pyrethrum
* Sesame oil
* Sulfur

Biological Controls see *Controls: Section 3*

Bacteria:
* ✱ *Saccharopolyspora spinosa* (Spinosad)

Bugs:
* minute pirate bugs *(Orius spp.)*

Fungus:
* ✱ *Beauveria bassiana*

Midges, Gall:
* ✱ *Feltiella acarisuga* (Cecidomyiid gall midge)

Predatory mites:
* *Galendromus occidentalis*
* *Mesoseiulus longipes*
* *Neoseiulus barkeri* (or *Amblyseius barkeri*)
* *Neoseiulus californicus*
* *Neoseiulus fallacis* (*Amblyseius fallacis*)

Phytoseiid mites:
* *Phytoseiulus persimilis*

Thrips

It's a sultry summer afternoon. The air is warm and heavy with moisture. On the horizon, an approaching thunderstorm rumbles. The thunderbugs rise up in a swarm, attacking plants, grasses, and those unfortunate enough to be outside. The thunderbugs are thrips, tiny insects ranging in size from about 1/16 to 1/5 of an inch. They are nicknamed thunderbugs or thunderflies because they become especially active during thunderstorms.

Thrips are so small that they must be magnified to see them clearly, but they are the only marijuana-eating insect that that is likely to invade your home and that might bite you.

The term thrips is both singular and plural. Thrips are members of the order Thysanoptera, which in Greek means fringed (*thysanos*) and winged (*pteron*). These slender insects have long fringes on the edges of their two pairs of narrow wings. They range in color from translucent white to black. Immature thrips, called nymphs, are similarly colored, but wingless.

There are more than 5,000 species of thrips. Their feeding preferences vary. Some are herbivores, others are predators, and some are omnivores: They eat everything. They are all poor fliers and usually jump when disturbed. Aloft, they rely on breezes and wind to compensate for poor flying ability.

They sometimes swarm. Clouds of them are easily sucked into unfiltered ventilation systems and open vents. Swarms of thrips enter buildings and cover surface areas such as furniture, walls, and curtains. They are attracted to white walls and curtains, but blue and pink are the colors they find most attractive. Swarming must have a purpose, but we still have only theories. One is that swarming relates to finding mates. Another is that

At a glance

Symptoms

Swarms of miniscule insects that become more active during stormy weather; leaves whiten, are scarred, and appear scabby with dark feces scattered over the leaves; heavily infested leaves look like the chlorophyll has been sucked out. Thrips are vectors for viruses that cause other symptoms: wrinkled leaf margins and crinkled leaves, rolled up leaves and leaf tips, yellow chevron strips, or light-green halo formations on leaves.

Actual size

1/15" • 1.5 mm

Leaves are speckled with white spots, indicating damage.

they detect changes in the atmosphere foretelling storms and head for cover.

Some species, including flower thrips, are likely to pinch your skin using their mandibular stylets. The attack does not break the skin and usually has no lasting effect. However, it is slightly painful and some people develop a small skin irritation.

Thrips pose two threats to plants. First, they scrape the leaves to obtain food and then they drink the juices, weakening or even killing the plant. Secondly, some species are vectors for viruses that affect marijuana. As the tiny pests feed off the wounded tissue, they transmit viruses, resulting in decreased vigor, lower yield, or even death.

Thrips attack indoor and greenhouse gardens, and outdoor gardens in temperate regions. Commercial greenhouse operators find them fairly easy to control but very difficult to eliminate.

Thrips prefer areas of rapid growth so they are usually found on young leaves, flowers, and terminal buds. Thrips' damage is usually detected before they are. Their piercing and rasping technique of feeding results in white scars and scabs on the leaves.

Thrips' damage may be mistaken for the stippling damage of spider mites or leaf miners. However, on closer examination, thrips' damage looks like little scrapes while bug and mite damage appears as punctures.

Unlike bugs such as aphids or leafhoppers, which have two hollow mandibular stylets, thrips have only one, which is closed at the end, like a needle. They pierce plant tissues with it. Then the stylet sinks deeper into the plant and sucks up plant juices from the ruptured cells.

Thrips have two types of saliva that they use while feeding. One digests plant tissue, liquefy-ing it so that it can be ingested. The other inhibits the plant's ability to coagulate the flow of sap from the wound.

LIFE CYCLE

Thrips undergo gradual metamorphosis, which is between complete metamorphosis, like a butterfly, and incomplete metamorphosis, like a grasshopper. The life stages are egg, nymph, prepupa, pupa, and adult.

Almost all species you are likely to encounter are from the suborder Terebrantia, which are distinguished by the presence of an ovipositor. This structure on the end of the female's abdomen is used to slit the leaf tissue and place a single egg in the opening. Eggs hatch in seven to 20 days, depending on temperature. Thrips nymphs look

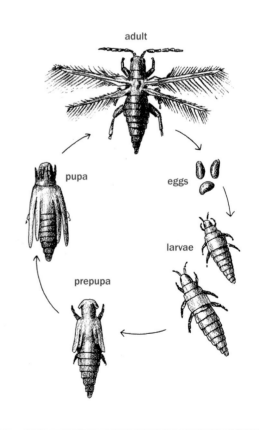

much like their parents except they are wingless. They feed voraciously and pass through two molts. The third and fourth instars are called prepupae and pupae. Nymphs do not actually pupate, like butterflies. Instead, they seek out dark crevices in the plants or drop off the plant and burrow into the soil or leaf litter where they remain immobile. This stage lasts about a day.

Some species form a pupa or cocoon. Thrips in the suborder Terebrantia pass through a single pupal instar during which the wings grow and the insects mature into their adult form. The life cycle is completed in 30 to 45 days depending on temperature and food quality.

Thrips can reproduce parthenogenically, or without the use of males. That is the preferred method with most species. Males are rare in most populations. In some species, only females have been observed. Virgin females produce only females. Mated females produce both males and females.

Outdoors, thrips overwinter in the soil and plant debris as pupae or adults. With the approach of spring, they become active when temperatures reach 60 °F. The amount of damage they cause is determined by temperature—the pests become more active and cause more damage as it increases. Indoors, many species, including the glasshouse thrips, remain active all year.

Marijuana plants are most likely to be attacked by the following thrips:

Frankliniella occidentalis (Western Flower Thrips)

Western flower thrips were confined west of the Mississippi River prior to 1980. Today, they are found all over the United States as a result of contamination of stock from California and by floating on wind current. They were found in Holland in 1983, and infested most of Europe shortly after. Western flower thrips are serious pests in greenhouses, infecting more than 250 species of plants, and are vectors for many viruses.

Adult females are about 1.3 to 1.7 millimeters in length. They vary in color from pale yellow to dark brown with dark bands across the abdomen. Their heads are lighter than their bodies and their eyes are dark. Forewings are light colored as well. Males are smaller, about 1.1 millimeters in

Table: Thrips Species				
Species Name	**Common Name**	**Disease Vector**	**Ovipositor**	**Bites Humans**
Frankliniella occidentalis	Western flower thrips	Y	Y	N
Frankliniella tritici	Flower thrips	Y	Y	Y
Heliothrips haemorrhoidalis	Greenhouse thrips	N	Y	N
Hercinothrips femoralis	Banded greenhouse thrips	N	Y	N
Oxythrips cannabensis	Marijuana thrips	Y	Y	?
Thrips tabaci	Onion or tobacco thrips	Y	Y	N

length, and are uniformly yellow. They resemble the eastern flower thrips, but their bodies are darker. Females prefer flower buds and flowers.

Females lay from 150 to 300 eggs over their lifetimes. Eggs are yellow but are not visible because they are buried deep within plant tissue. They hatch in two to 14 days depending on temperature. Nymphs start out yellow, but turn whitish as they approach pupation. Yellowish prepupae are found on leaves then drop into the soil for the final phase. The life cycle takes from 12 days in warm temperatures to 45 days under a cooler regimen. Optimal temperatures are between 79 to 85 °F.

Frankliniella tritici (Flower Thrips)

The eastern flower thrips is a biter, and on occasion attacks humans. They are most likely to be a pest in greenhouses and commercial floriculture operations. They swarm in large numbers on warm afternoons and they have been captured at altitudes of 10,000 feet. Besides biting the hand that feeds them and damaging buds and flowers, flower thrips are also vectors for a number of diseases, including Tospoviruses.

This thrips was once dominant east of the Rocky Mountains, but has been replaced by the western flower thrips. *F. tritici* prefers the most tender, newest growth of grasses and light-colored or yellow blossoms. Their heavy feeding can stunt and disfigure the host.

Adult females are about 1.25 millimeters in length and are yellow with brown blotching. Males are smaller and lighter in color. Females lay kidney-shaped, pale yellow eggs buried deep in the plant tissue of flowers, buds, or furled tissue. Nymphs are lemon yellow with no wings. There are three nymph instars. The pupae drop to the soil or leaf litter. New adult females begin laying eggs one to four days after emergence in the summer and 10 to 15 days in cool winter greenhouses. The life cycle takes from 10 to 35 days depending on temperature. Adults overwinter outdoors in grass clumps and other sheltered places, as far north as North Dakota.

Thrips nymphs.

Heliothrips femoralis
(Banded Greenhouse Thrips)

Banded greenhouse thrips feed on a variety of plant hosts from cucumbers to screwpine to moonflowers. In a test of 50 randomly offered hosts, *H. femoralis* infested 44 of them.

Banded greenhouse thrips nymphs are yellow-white and range from .5 to 1.5 millimeters long, wingless, and have red eyes. Colonies of nymphs congregate together to feed, encased in a water globule of excrement. The nymph stage lasts about 18 days, then the insects pupate. Prepupae and pupae are white and remain on the leaves. Adults emerge shortly thereafter and feed less ravenously than the nymphs. Female banded greenhouse thrips are approximately 1.5 millimeters long with red eyes and gray-brown narrow fringed wings with three white crossbands. They are primarily yellow at first but gradually darken to brown or black. Males are rare.

Sometimes referred to as sugar beet thrips, *H. femoralis* has gained a well-earned reputation as a serious floriculture and greenhouse pest.

The translucent, elongated eggs are about 0.25 millimeters long and are deposited on the undersides of leaves or along the stems. Eggs hatch in two weeks. Adults live about 40 days, more in cool temperatures, less under hot conditions.

Heliothrips haemorrhoidalis
(Greenhouse Thrips)

Greenhouse thrips are not host specific. They are found in tropical environments and in greenhouses everywhere. Native to Central and South America, they were introduced to Europe, Africa, Asia, and North America on tropical plants bound for greenhouses.

Adult females are 1.3 to 1.8 millimeters in length, black with whitish-to-yellow legs and translucent wings folded back over the thorax and abdomen. Males are slightly smaller and wingless. Antennae have eight segments. Nymphs are white or yellowish with red eyes and no wings. They mature to about 1 millimeter. A droplet of fecal liquid forms at the end of the abdomen as the nymph feeds, then drops off onto foliage, and another droplet takes its place. These droplets dry into dark flecks of fecal matter on leaves and greenhouse surfaces. This sticky droplet also protects the nymph by gumming up parasitoids and repelling predators.

The prepupae are a light yellow with red eyes. The pupae start out yellow and turn brown. Their eyes are larger than the prepupae and the antennae bend backwards over the head.

The life cycle for *H. haemorrhoidalis* is about 30 to 33 days at 78 to 82 °F. Warmer temperatures mean more shorter life cycles per year. Females lay about 45 banana-shaped, white eggs in plant tissue, leaving a tip sticking out. Most reproduction is parthenogenic, so most of the young are females. Males are rarely seen. Greenhouse thrips pupate on the plant. *H. haemorrhoidalis* does not prefer marijuana as a host, but attacks it if there aren't other plants around.

Oxythrips cannabensis
(Marijuana Thrips)

Marijuana thrips are host-specific on cannabis, but are rarely encountered. They have been collected from the leaves and flowers of feral hemp in Illinois and were isolated from hemp in the Czech Republic and Siberia. *O. cannabensis* infests the leaves and flowers and has been considered as a biological control against illegal marijuana. It is thought to be a vector for plant pathogens, but little is presently known of its life history.

Thrips tabaci (Onion or Tobacco Thrips)

Adult female onion or tobacco thrips are yellow with brownish blotches on the thorax. Males are rare and sexual reproduction almost never happens, resulting in a population of almost all females. Females are about 1.2 millimeters in length and lay 20 to 200 very small, white, kidney-shaped eggs deep inside plant tissue using their ovipositors. Eggs hatch in six to eight days and the first instar nymphs are white, 0.35 to 0.38 millimeter in length. Second instar nymphs are yellow and a bit larger. Nymphs pupate in about 14 days and emerge as adults in another five to nine days. Adults and pupae overwinter in the soil or plant debris. The life cycle is about 20 days at higher temperatures, but at temperatures of 60 °F, life cycles elongate to 35 days. Onion thrips attack a wide variety of plants and are the thrips most likely to transmit viruses, including the hemp streak virus, the cannabis pathogen Argentine sunflower virus, and other diseases due to their feeding by piercing the plant tissues.

Virus Transmission

There is no cure for plant viruses. Infected plants should be removed and gardeners should wash themselves and their clothes before they reenter the space.

Tospovirus Infection Symptoms

Ringspots, which can be white, yellow, or tan appear around the site of the thrips penetration. New leaves are distorted and crinkled. Growth is stunted or one-sided.

Hemp Streak Virus

Chlorotic areas develop in a series of interveinal yellow streaks or chevron stripes. Brown, necrotic flecks surrounded by a pale green halo may develop. Flecks most often appear along the margins and tips of older leaves. Streak symptoms are most likely to develop in moist weather. Fleck symptoms most likely develop during dry conditions. Eventually leaf margins wrinkle, leaf tips roll upward, and leaflets curl into spirals.

CONTROLS
Thrips Detection

The rasping and puncturing pattern of feeding used by thrips results in the characteristic damage that identifies an infestation. Plant tissue injured by thrips heals, leaving small, irregularly shaped, and elongated white or yellow lesions in between the veins. Leaves are dotted with these lesions and specks of thrips' fecal matter. Both the lesions and poop are seen mostly on leaf undersides but also on the top of the leaves. Detection and control of infected plants is a two-step process. Inspect living plant leaves on a regular basis and set out detection traps.

Blowing on blossoms and growing plants causes thrips to become mobile and spring away to a safer location. They are reacting to the CO_2 in your breath and to the disturbance of the foliage.

The second step is to set out sticky traps. Yellow sticky cards, which are used for fungus gnat detection, attract thrips, but they prefer blue and hot pink cards. You can make traps by smearing petroleum jelly onto suitably colored cardstock or Post-Its.

Colored cards are more useful as simple indicators than as a form of control.

Infestation Severity Index for Thrips	
Light	Any thrips damage seen
Moderate	Thrips damage and occasional thrips seen
Heavy	Thrips damage on many plants but confined to lower leaves OR two to 10 thrips per leaf
Critical	Thrips damage on growing shoots OR greater than 10 thrips per leaf

Table found in *Hemp Diseases and Pests*, J.M. McPartland.

Barriers

One way to discourage thrips from taking up residence in an indoor garden or greenhouse is by creating a 10-foot dead zone around it to eliminate areas where thrips can live and breed. Remove plant debris from the garden, the growing area, and from greenhouse benches to eliminate locations where thrips may hide.

Thrips are likely to infest through the ventilation system. For this reason air-intake barriers are extremely important in indoor grows and greenhouses. Not all barriers are effective because thrips are so tiny. HEPA filters and the insect barrier products Fly-Barr, BugBed 123, and No-Thrips are designed to exclude airborne thrips.

You may have noticed in descriptions of the different thrips species that the larvae drop down to the planting medium to pupate. They spend this transitional portion of their lives underground in the soil, from which they will emerge as adults. This is a great time to interrupt their life cycle. If the larvae cannot reach the soil, they will just shrivel up and die after they drop down onto a barrier. Barriers can easily be made from paper, cardboard, or other materials. Coir pot covers are also available. With the life cycle interrupted, in two to three weeks old thrips die and the infestation ends.

For greenhouse or indoor planting areas, a short fallow period can be used to eliminate an infestation. Remove all plants and then heat the garden space. The thrips' eggs hatch but have no food so they die of starvation. To totally eliminate thrips the process takes about a week.

Pesticides

Garlic sprays repel thrips, and outdoors, garlic plants can be used as companion plants. Adding capsaicin increases the effectiveness. Pyrethrum is effective but must be used several times to eliminate the population because it breaks down over a few days in the presence of light and air. Neem oil, a multipurpose, is most effective on thrips when they are in the larval phase.

Herbal oil-based pesticides are effective because they attack thrips by causing deterioration of their tissues. Insecticidal soap such as Safer Brand smothers and coats the spiracles—the breathing holes—found all along the insect's body.

Sierra Natural Science's SNS-203, made from rosemary and clove oils, works on soil-based pests.

Biological Controls

Thrips do have some predators, and these insects can be used as beneficials for indoor gardens or greenhouses. The bacterium spinosad is also effective

BUGS

Six different species of minute pirate bugs are available for thrips control.

Adults eat five to 20 thrips per day. Once the thrips population is under control, pirate bugs survive on aphids, mites, scales, whiteflies, other biological controls, and each other. They eat the largest pests first and fly away if food becomes scarce. They prefer flowering plants. All species are sensitive to pesticides.

FUNGUS

Beauveria bassiana: This beneficial fungus can work to control many pests, including thrips.

Adding neem oil increases the effectiveness.

MITES, PREDATORY

H. miles: The *H. miles* mite preys on thrips pupae, but also likes fungus gnat larvae and other soil insects. *H. miles* is compatible with *B. bassiana* and *Steinernema* species of beneficial nematodes.

Iphiseius (Amblyseius) degenerans: This mite attacks the first larval stage of onion (*T. tabaci*) and western flower thrips (*F. occidentalis*) and also eats aphids, spider mites, whitefly eggs, small caterpillars, and pollen. It remains active under the short-day lighting commonly used in flowering rooms. This beneficial tolerates neem oil and fungicides, but is sensitive to insecticides, especially soap, sulfur, and pirimicarb.

Neoseiulus (Amblyseius) barkeri: This mite preys on onion thrips (*T. tabaci*), *F. occidentalis,* and spider mites. It enters diapause under short photoperiods.

Neoseiulus (Amblyseius) cucumeris: This predatory mite feeds on thrips' eggs and the first instar, and also eats mites and pollen. It is good for vegetative rooms because it prefers foliage rather than flowers, doing best in humid conditions. It is not very effective in flowering rooms because it goes into diapause under short-day regimes.

Phytoseiid mites: These mites are generalists, eating thrips, aphids, and spider mites. They prefer two-spotted spider mites, but will feed on thrips, honeydew, and pollen if preferred prey is minimal.

NEMATODES, BENEFICIAL

Several beneficial soil nematodes will control thrips species that pupate in the soil.

Heterorhabditis bacteriophora: This nematode tracks down hosts to parasitize by following their CO_2 and feces trails. They cannot reproduce within thrips' cadavers like they can in some other species of hosts, so repeat applications are necessary.

Steinernema feltiae and *Steinernema riobravis* (corn earworm) also parasitize thrips pupae, but *H. bacteriophora* is by far the better control.

WASPS, PREDATORY

Ceranisus menes: This wasp parasitizes *F. occidentalis* and *T. tabaci* larvae.

Eulophidae parasitoids: Two species of Eulophid wasps are available commercially for thrips control.

Thripsobius semiluteus: This tiny wasp parasitizes greenhouse thrips (*H. haemorrhoidalis*). Do not use insecticides within 30 days of release.

Trichogrammatidae parasitoids, also known as chalcid wasps, are parasites of a lot of different insect pests. Some species of Trichogrammatidae in the *Megaphragma* genus parasitize western flower thrips.

SUMMARY
Thrips Controls

Look in Part 2: Controls for more detailed explanations of the solutions listed below. Part 2 is divided into four sections, each representing different strategies.

Preferred methods are marked with an ✱ asterisk.

Preventive Measures

Use air-intake filters, herbal oils, and the biological control team of nematodes, minute pirate bugs, and *Beauveria bassiana*. Prevent larvae from reaching soil with barriers. Remove plant debris from the garden, the growing area, and from greenhouse benches to eliminate locations where thrips can hibernate.

Allow a fallow period, during which thrips and their eggs will perish.

Barriers & Physical Controls see *Controls: Section 1*
* ✱ Air-intake filters
* Aluminum foil
* Heat
* ✱ Sticky cards (for detection)

Pesticides see *Controls: Section 2*
* Capsaicin
* Garlic
* ✱ Herbal oils
* Insecticidal soap
* Neem oil
* Pyrethrum

Biological Controls see *Controls: Section 3*
Bacteria:
* ✱ spinosad
Beetles:
* lady beetles
* Rove beetles (*Atheta coriaria*)
Bugs:
* ✱ minute pirate bugs (*Orius* sp.)
* mirid bugs
Fungi:
* ✱ *Beauveria bassiana*
* ✱ *Verticillium lecanii*
Predatory mites:
* ✱ *Hypoaspis miles*
* *Iphiseius (Amblyseius) degenerans*
* *Neoseiulus (Amblyseius) barkeri*

* ✱ *Neoseiulus (Amblyseius) cucumeris*
Beneficial nematodes:
* ✱ *Heterorhabditis bacteriophora*
* ✱ *Steinernema feltiae*
Predatory wasps:
* *Ceranisus menes*
* Eulophidae parasitoids
* *Thripsobius semiluteus*
* ✱ Trichogrammatidae parasitoids (chalcid wasps)

Tobacco Mosaic Virus

At a glance

Symptoms
Plant leaves develop splotchy, mottled, multicolored spots that have an appearance similar to a mosaic or paisley pattern.

Infected plants.

Incurable, untreatable, contagious, and able to infect even after being dried or frozen, the tobacco mosaic virus is one of the most devastating garden problems to encounter. The most obvious indication of infection is a plant having distorted leaves with paisley-calico patterns of light green, dark green, yellow, and possibly brown splotches.

Over 100 types of herbaceous plants are susceptible to members of the tobacco mosaic virus (TMV) family. Hemp mosaic virus (*Comoviridae comovirus*) is the name given to the strain of the tobacco mosaic virus that is most likely to infect cannabis. Other names for this strain of the virus include cowpea yellow mosaic virus, crotalaria mucronata mosaic virus, and Sunn-hemp rosette virus.

In cannabis, hemp mosaic virus first appears as light green spots on plant leaves between the leaf veins. This virus is a rod-shaped piece of ribonucleic acid (RNA) with a protective protein coating. These submicroscopic viruses are only 300 nanometers (0.000012 inch) long.

Once introduced through a wound or insect bite, the virus will pass into individual plant cells. Inside a cell, it will shed its protein coating, and take control of the cell to replicate itself. This process of making copies of itself interferes with the normal functioning of the cell, which is what makes the plant sick. As the virus reproduces, the offspring moves into other plant cells and the infection spreads. As the virus spreads through the plant the number of these malfunctioning cells increase until groups are visible on the plant leaves. As it progresses, the spots get larger and develop brown rings. The leaves distort, and raised areas may appear. These areas of distorted cells form its namesake mosaic pattern.

Although the virus does not produce spores to infect nearby plants like a mold, it will transmit to healthy plants by touch, insect carrier, or pollen release. Hands and tools can become contaminated with the virus and then infect healthy plants they come into contact with. In greenhouse or field conditions where plants are kept close enough to touch, hemp mosaic virus can spread rapidly. Insects can transfer the disease by feeding on an infected plant and then traveling to an uninfected plant. If an infected male plant is allowed to flower, the pollen can spread the virus to healthy plants. Any seeds produced from an infected parent may carry the infection as well, and if sprouted the offspring may develop the virus.

Unlike many other viruses, it retains its ability to infect growing plants long after the death of the host plant. Dried and cured buds from a contaminated plant can infect growing plants if they come into contact. For cannabis, the destruction of the infected host and sterilization of the area is the safest treatment. Tools or hands that have been in contact with infected plants should be cleaned after use to prevent spreading the disease. Use a strong soap to clean skin. Tools can be soaked in a disinfectant. For infected material able to withstand the temperature, heat sterilization takes place at a temperature of 200 °F at 45 minutes to an hour.

CONTROLS

Once infected, there is no cure for hemp mosaic virus—plants may be able to survive until a reduced harvest is collected, but growth will be stunted, and the final product will be of inferior quality.

Infected plants and media should be removed and quarantined or disposed of as quickly as possible as the virus is communicable to uninfected healthy plants via insect bite or wound exposure to the virus. Consider any clones recently taken from an infected plant to be suspect as well, and destroy them. If left unchecked, hemp mosaic virus can ruin an entire cannabis garden.

The key to controlling hemp mosaic virus is preventing the initial infection. Proper sanitation and preparation as described in Preventive Maintenance can help mitigate outbreaks. As chewing insects are a contamination vector, the controls outlined in Part 2: Controls can help prevent outbreaks.

Commercial cigarette tobacco may be contaminated with the virus and therefore should be prevented from coming into contact with cannabis plants. For this reason, tobacco consumption in close proximity to growing cannabis plants is discouraged. Homemade nicotine-based insecticides made with infected tobacco may transmit hemp mosaic virus to sprayed plants unless boiled for 20 to 30 minutes.

Once an infection has taken place, the virus can withstand exposure to winter temperatures or summer temperatures over 100 °F. It has been known to survive in dried plant material for half a century or more, and if frozen can stay viable indefinitely.

In an outdoor setting, not only should the infected plant be removed, but also any plants within root range of it. Roots left in the soil after the diseased plant has been removed can be sources of infection. One of the most common infection sources is plant material left in the soil after the removal of an infected plant. Do not immediately replant exposed soil with any plant susceptible to the virus.

Problems with additives, herbicides, nutrients, and air pollution can all mimic the effects of the hemp mosaic virus. As similar symptoms can be caused by a variety of growing issues, do not be hasty in condemning a suspect plant to destruction. If feasible, one option is to quarantine and rule out curable problems before destroying the host.

Fortunately, all forms of the tobacco mosaic virus, including the hemp mosaic virus, are harmless to humans even if consumed.

If you decide to try and slow down the development of the virus in an infected plant long enough to harvest, treatment must be implemented when the symptoms first become apparent. The virus cannot be cured, but the progress can sometimes be slowed down long enough to salvage a reduced harvest.

Boosting the plant's systemic acquired resistance response by treating with aspirin can help a diseased plant. Aspirin mixed at 250 to 500 milligrams (one or two tablets) per gallon of water can be used as a drench or a foliar spray. Willow water made from soaking willow bark in water overnight can also be used for a similar effect.

These steps in keeping your plants alive longer should only be taken if absolutely necessary, as the longer the plant lives, the more chance it has to infect other plants; they are still contagious even with treatment. Removal and disposal of the infected plants is generally recommended.

Light green coloration radiating from veins.

Mottled mosaic pattern appears on leaves.

Leaves may exhibit odd growth such as twisting, stretching, and crinkling.

Tobacco (Hemp) Mosaic Virus Controls

Look in Part 2: Controls for more detailed explanations of the solutions listed below. Part 2 is divided into four sections, each representing different strategies.

Preferred methods are marked with an ∗ asterisk.

Preventive Measures

Limit plant exposure to tobacco products, including cigarette smoke. Wash hands after smoking cigarettes. Removal and destruction of the infected plant is recommended, but if impractical to do so, treat with aspirin.

Barriers & Physical Controls see *Controls: Section 1*

Removal of infected plants

Outdoor Strategies see *Controls: Section 4*

Remove weeds or other plants from the area that may serve as hosts. Ideally, a non-susceptible crop should be grown in infected ground for two years after infection.

Verticillium Wilt

At a glance

Symptoms
Starting at the bottom, parts of the leaves turn yellow. This spreads to whole leaves, then they turn brown. At the same time tan or brown striations appear lengthwise along the stem. There is brownish discoloration between the pith and the woody part of the stem. Shortly, the plant wilts and dies.

Infected hops leaf shows typical symptoms.

Verticillium wilt is one of the nastier fungi in the garden. There is no cure or treatment for it. It is not a common problem in cannabis gardens, but it is one of the most destructive. Verticillium wilt can ruin not only the current crop, but future harvests in the same location unless the area is cleaned of spores.

The most obvious indication of an infection is yellowing leaves and browning and wilting stems. The fungus grows into the vascular system, blocking nutrients and water flow through the plant. After the leaves wilt, they will drop, and the plant will die.

Verticillium wilt enters plants through the root system. Once a root becomes infected, the fungus spreads by growing inside the plant, and by releasing spores into the vascular system that are transported through the plant quickly. Because of this systemic means of infection, only some of the branches may be infected while others remain healthy.

Infected roots in contact with roots from a neighboring plant spread the fungus. A small infection often spreads this way infecting a large area in a short time. Aside from cannabis, Verticillium wilt attacks a wide variety of herbaceous and woody plants, including peppers, melons, strawberries, and tomatoes, as well as many fruit and shade trees. The same fungus is infectious across crops.

Before diagnosing a plant with Verticillium wilt, make sure that the plant does not have a nitrogen deficiency coupled with a lack of water, which can have a similar appearance. If the plant doesn't respond to watering by perking up, then Verticillium wilt is the culprit. To verify infection, split the stems and trunk to look for a brownish or dark discoloration between the edge of the pith (the white part) and the outside edge.

PREVENTION

Verticillium wilt is a very virile fungus that is difficult to kill. It can survive being frozen or dehydrated. Some report that in infected soil it can remain viable for a decade or longer. Treatment for infected soil includes soil solarization, flooding, and crop rotation with resistant plants for three to five years.

Rotate crops. Avoid planting the same crop in the same ground for many years in a row. Even if none of the plants show symptoms, multiple successive plantings may cause the fungus to build up in the soil until it reaches destructive levels.

Compost tea is a gardeners' aid in preventing Verticillium wilt. Microbes in compost tea are antagonistic to Verticillium fungi and help reduce infections.

Mycorrhizae fight pathogens and increase plants' disease resistance.

Inoculate soil with beneficial fungi and bacteria before planting and inoculate plant roots as well, using a root dip or irrigation.

CONTROL

A Verticillium wilt outbreak must be dealt with harshly. Any infected plant material, growing media, and any suspected contaminated material should be carefully removed from the garden and disposed of. Verticillium wilt spores can remain viable in soil for many years. Infected fields are generally considered to no longer be suitable growing sites for non-resistant plants.

All tools, pots, trays, and items that have come into contact with the infected plant should be thoroughly cleaned or disposed of.

Verticillium Wilt Controls

Look in Part 2: Controls for more detailed explanations of the solutions listed below. Part 2 is divided into four sections, each representing different strategies.

Preferred methods are marked with an * asterisk.

Preventive Measures

Mycorrhizae and compost tea applications help to prevent initial outbreaks. A plant infected with Verticillium wilt is a danger to the rest of the garden, and a single infected crop can cause problems for several years if the soil is not treated or removed. Carefully remove and destroy any infected plants.

Barriers & Physical Controls see *Controls: Section 1*

- Physical removal of infected plants

Fungicides see *Controls: Section 2*

- * Compost tea
- * Seaweed powder or liquid

Biological Controls see *Controls: Section 3*

- * Bacteria:
  - *Bacillus subtilis*
  - *Streptomyces griseoviridis*
  - *Streptomyces lydicus*
- * Fungus:
  - mycorrhizae
  - *Pythium oligandrum*
  - *Trichoderma* species, including *T. harzianum*

Outdoor Strategies see *Controls: Section 4*

Crop rotation: Ideally, a non-susceptible crop should be grown in infected ground for three to five years after infection.

Flooding: Verticillium wilt can be killed if submerged for two weeks.

Remove weeds or other plants that may serve as hosts from the area.

Soil solarization: Soil may be infected deeper than is normally heat-treated, so some excavation may be required.

Whiteflies

On first notice, an infestation of whiteflies seems almost pretty. These tiny white insects flutter up delicately when their host is disturbed, creating the image of a tiny snowstorm. This idyllic notion dissipates quickly for anyone who has ever dealt with these soft-bodied sucking machines. Whiteflies are prolific pests that make a banquet out of the garden, often inviting other types of pests along to the party.

Whiteflies are not really flies at all. Nor are they moths. They belong to the order Hemiptera, which means "same wing." Like all Hemiptera, they are "true bugs." They have two sets of wings, front and back, that are identical, and this characteristic qualifies them for membership in this order, along with their cousins leafhoppers, mealybugs, aphids, and scales. They are tiny, about 1/16-inch long and usually appear in great numbers. Where there's one, there are always more.

Whiteflies derive their name from the white, waxy substance that coats their bodies and wings that is produced by special glands in the whiteflies' abdomens. They rub this material over their bodies, presumably to help keep them from drying out. They also secrete a waxy substance during the immature nymph stage.

Whiteflies are likely to be noticed before there are signs of their damage. When vegetation they are sucking on is shaken they flutter off. Their tiny white bodies are very noticeable. Then they are likely to fly back to the plant they were on.

The symptoms of a whitefly infestation are subtle at first—a few tiny, yellowish leaf spots here and there, that can easily be mistaken for aphid damage. Mistakenly identifying whiteflies can allow a low-level infestation to quickly balloon into an all-out invasion.

At a glance

Symptoms
Small white insects that flutter close to plants when foliage is disturbed. The tiny, yellow-brown leaf spots are easily mistaken for aphid or mite damage; undersides of leaves have circular or patterned bumps of eggs that may be coated with a waxy substance; tiny clear empty pupae casings split in T-shaped pattern litter underside of leaves.

Actual size

1/16" • 1.5 mm

Whiteflies pierce the leaf with their proboscises.

Whiteflies lay eggs on the undersides of leaves in circular or random patterns, which they coat with a waxy, filamentous substance. Another early sign of whitefly infestation are the cast-off shells of the pupae—tiny, clear casings on the undersides of the leaves that are split in a distinct T-shaped pattern.

LIFE AS A WHITEFLY

Whiteflies are the bane of indoor and greenhouse growers. In the indoor space or greenhouse, when the whitefly life cycle proceeds uninterrupted, these bugs reproduce exponentially until the air is like the inside of a snow globe when they fly, and they've coated the undersides of leaves with their larvae and cast-off pupal cases

Whiteflies are sucking insects. They pierce the tissues of plants and suck out the phloem juices, resulting in wilted, stunted, and sometimes dead plants. However, sucking juices often isn't the worst problem whiteflies create. They also transmit diseases to their hosts. When they pierce a plant with their mouthparts, or "rostrums," whiteflies inject saliva into the plant, which passes along viruses or bacteria to the plants. Whiteflies transmit a number of viruses and bacteria, including hemp streak virus, which is also transmitted by thrips.

Whiteflies produce a waste substance called honeydew that can cause even more problems. Honeydew, a sugar-rich liquid, is created when insects suck plant sap to obtain protein. Most of it is secreted from the tip of the abdomen as a slightly viscous substance that drops onto foliage, greenhouse benches, walkways, or mulch, generally creating a mess and attracting other pests.

Ants collect and feed on honeydew. Once they discover an abundant source, they take up

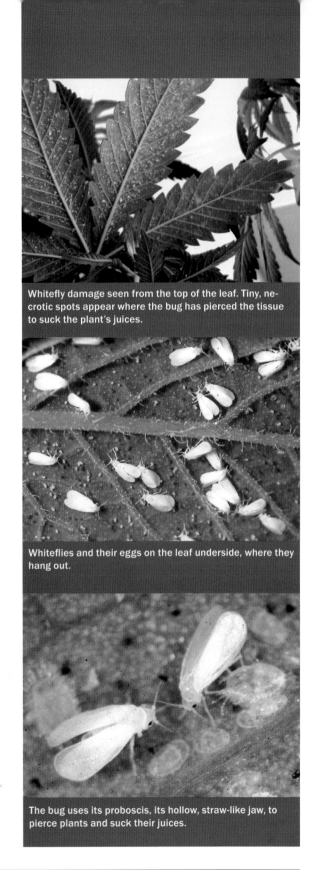

Whitefly damage seen from the top of the leaf. Tiny, necrotic spots appear where the bug has pierced the tissue to suck the plant's juices.

Whiteflies and their eggs on the leaf underside, where they hang out.

The bug uses its proboscis, its hollow, straw-like jaw, to pierce plants and suck their juices.

residence and may bring along friends, namely their own herd of honeydew-producing aphids or mealybugs—visitors you do not want in your garden. Ants defend the aphids, mealybugs, and whiteflies against predators including some beneficial insects introduced for control.

In addition to pests, the presence of honeydew encourages the development of mold. Sooty mold is a name used to describe several fungi that grow on honeydew. The specific combination of fungi that form sooty mold depends on the environment, the species of insect that produced the honeydew, and the host. As the molds grow on the nutrient-rich honeydew, the dark mycelia of the fungi gives the surface the appearance of being covered with soot. Sooty molds do not infect plants, but substantial growth on leaves blocks light from penetrating to the leaf surface, stifling photosynthesis, the process by which plants produce food. This seriously compromises the plants' health and can lead to their death.

The Life Cycle

Whitefly adults have short but prolific lives. They only live 21 to 36 days, mostly depending on temperature, but a female can lay between 60 and 100 eggs in that time.

Whiteflies reproduce either by mating or parthenogenically. Unmated females lay unfertilized eggs that hatch into males. After mating, the eggs hatch into females.

All species of whiteflies have similar life cycles. Depending on species, females deposit white eggs on the undersides of leaves in a random pattern, an arc, spiral, or circular pattern. Then they cover them with a white waxy substance.

After seven to 10 days, the eggs hatch into crawlers. This stage is the only truly mobile

stage of development until adulthood. Crawlers are between 0.5 to 1.5 millimeters in length and barely visible to the naked eye. They wander around on the underside of their leafy home until they find a suitable plant vein. Then, they sink their mouthparts into this vein and begin sucking out the juices. Once attached, they stay in this spot until they mature into adults.

As these crawlers dine on plant sap, they grow and change, passing through two more feeding stages as sessile nymphs. These nymphs are called sessile because the term means that they do not move. The nymphs are oval and have very simple antennae and legs, usually consisting of one segment, if the appendages are even visible. Whitefly species that attack marijuana tend to be yellowish-white in color. Other species that favor woody plants are dark tan, brown, or black.

In the next stage of development, nymphs

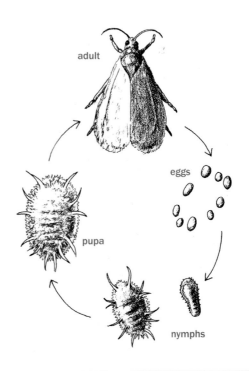

metamorphose into "pseudopupae." This is not a true pupal stage as in butterflies or moths. Instead, the nymphs lose most of their larval appearance and start to resemble an oval casing clinging to the leaf. It is at this stage that entomologists can first distinguish between species using a microscope to compare characteristic structures such as pores, wax filaments, spines, hairs, and grooves. Adults of most whitefly species develop more obvious differences, making it easier for the layperson to differentiate between them.

As the pupa matures, the outline of the yellowish, soon-to-be adult whitefly can be seen through the clear pupal case, or the shell that contains it. In some species, distinctive adult features begin to develop, such as the red eye spots found in the silverleaf whitefly. After about six days, the whiteflies mature and force the pupal case open in a T-shaped pattern. The new adults emerge and dry their wings before flitting away.

WHITEFLIES ON CANNABIS

Three species of whiteflies are reported as pests on cannabis.

Greenhouse Whitefly (*Trialeurodes vaporariorum*)

The greenhouse whitefly is the most common whitefly found in indoor and greenhouse growing. These tiny insects have an uncanny ability to find hosts, even indoor crops, that are supposedly "safe" from pests. Females lay more than 100 eggs on the undersides of leaves, positioning them in a circle or semicircle and attaching them with short stems. The eggs when first laid are green to pale yellow in color, but they mature to purple-gray or a brownish-black before they hatch into

the first crawler larval stage seven to 10 days later.

Greenhouse whitefly crawlers are transparent, oval, and flat with a row of waxy threads around the perimeter of their bodies. The crawlers begin to feed and quickly become sedentary and lose their legs. They are often mistaken for scale insects. The fourth stage instar secretes additional wax, producing a halo of more waxy threads around its edges. Adult emerge in two to four weeks depending on temperature; they prefer a temperature range between 70 and 90 °F. Adults are about 1 millimeter in length with four off-white wings held close over their backs, parallel to the leaf surface. Greenhouse whiteflies are vectors for hemp streak virus.

Sweet Potato or Tobacco Whitefly (*Bemisia tabaci*)

Sweet potato whiteflies are pests both indoors and outdoors. They prefer high humidity and temperatures ranging from the high 70s to high 80s °F. These whiteflies have a lifestyle similar to the greenhouse whitefly.

In outdoor gardens, sweet potato whiteflies overwinter as eggs, hatching in the spring. Indoors, generations overlap creating massive infestations if left untreated. Females, in their lifetime can lay 80 to over 400 eggs, which they hang from the leaf surface using short silken stems. Eggs mature from white to yellow before hatching into crawlers. As with the greenhouse whitefly, the larvae crawl around, choose a feeding spot, and then lose their legs.

The pupae of sweet potato whiteflies have reddish eye spots and wait out their pupal period in casings that taper to the leaf surface. They lack the waxy threads that halo the greenhouse whitefly pupae. Adults are yellow with angular rather

Differences Between Whitefly Species

	Greenhouse Whitefly	Sweet Potato Whitefly	Silverleaf Whitefly
Egg Color	green to pale yellow, then purple-gray to blackish-brown before hatching	white to yellow	white to yellow
Egg Placement	eggs laid in circles or semi-circles	eggs laid randomly over leaf surface	eggs laid randomly over leaf surface
Larvae & Puparia	edges of puparia perpendicular to leaf surface waxy spines on puparia	edges of puparia taper at 45-degree angle to leaf surface no waxy spines	edges of puparia taper at 45-degree angle to leaf surface no waxy spines
Adults	slightly larger, lighter colored, wings are rounded, hold wings closer to body	smaller, angular wings, hold wings tent-like over body	smaller, angular wings, hold wings tent-like over body

than rounded wing tips. Their wings, which they hold in a tent-like position over their backs, are marked with striations

Silverleaf Whitefly (*Bemisia argentifolii*)

The silverleaf whitefly appeared in the United States around 1986, when it was first identified in Florida as the "poinsettia whitefly." It has devastated both indoor and outdoor gardens in the southern states. Abundant rain has slowed its permanent colonization in some parts of the southeastern United States, but it is widespread in southern California, where it enjoys a diverse range of hosts and has evolved resistance to nearly all pesticides.

The life cycle of the silverleaf whitefly is identical to both the greenhouse whitefly and the sweet potato whitefly except that the females lay about 10 percent more eggs and produce more honeydew than the other two species.

CONTROLS

Whiteflies' presence in the garden can be detected using yellow sticky cards. They can be physically removed by vacuuming them off of the plants. This works to reduce the population because they hover around the plants. Morning is the best time because that is when they are slow moving. Spraying the crop with water regularly also reduces the population.

Outdoors, horticultural covers, which are mostly useful for protecting small plants, also prevent whiteflies from getting to the plants. Using aluminum foil as a mulch deters whiteflies from infesting small plants because it reflects UV light, which seems to confuse the pests.

Indoors, air-intake filters prevent whiteflies from entering indoor gardens through the ventilation system

One way of containing an infection is by removing heavily infested plants before any more eggs can hatch and spread. Do this during the cool part of the cycle to capture the pests when they are most sluggish. Cover the plant using a plastic bag, then remove it from the space.

Biological Controls

Whiteflies are particularly resistant to insecticides because of heavy pesticide use. Luckily, they have a number of natural enemies that can be utilized as biological control measures to eradicate or lessen an outbreak. They include parasitic wasps, fungi, and beetles.

Outdoors, the removal of natural enemies by heavy insecticide often results in a more robust population. With heavy infestation, using predators and parasites together produce better results. Below is a list of biological controls:

- The parasitic wasp *Encarsia formosa* is the most successful biocontrol organism for whitefly control. The greenhouse whitefly is its preferred host. A new race, the Nile Delta strain, attacks all major whitefly species so identifying the whitefly strain is not necessary.

E. formosa wasp using its ovipositer to lay an egg inside the whitefly. Upon hatching, it eats the whitefly larva from within.

- *Delphastus pusillus*, the beetle from the ladybug family, prefers whitefly eggs and pupae.

- The big-eyed bug (*Geocoris punctipes*).

- Damsel bugs (*Nabis alternates*).

- Mirid bugs (*Marcolophus caligenosus*).

- Minute pirate bugs (*Orius spp.*).

- Green lacewings (*Chrysoperla spp.*).

In addition, two types of fungi are effective for whiteflies: *Beauvaria bassiana*, which can be purchased as BotaniGard and Naturalis-O, and *Verticillium lecanii*. In addition, a newer class of biologically derived insecticides is based on the bacterium *Saccharopolyspora spinosa*, marketed as Spinosad. This bacterium produces insecticidal metabolites.

Pesticides

Pesticides are the least likely means of controlling whiteflies because they have developed a remarkable resistance to insecticides. Even so, there are a few pesticides you can try.

- Herbal sprays, made from essential plant oils, which may include many herbs and spices such as black pepper, cinnamon, clove, hot pepper, garlic, rosemary, and thyme. Whiteflies do not have resistance to these sprays because they deteriorate tissue upon contact.

- Neem oil can also be used but requires repeat applications because, like other botanicals, it does not persist on plant surfaces and breaks down in sunlight.

- Insecticidal soaps work on many soft-bodied insects, and can assist with reducing the whitefly population.

- Pyrethrum may be effective, but some white-fly populations are immune due to its lengthy use in pest control.

In outdoor settings, companion or trap plants can manipulate whiteflies into steering clear of the cannabis garden.

SUMMARY
Whitefly Controls

Look in Part 2: Controls for more detailed explanations of the solutions listed below. Part 2 is divided into four sections. each representing different strategies.

Preferred methods are marked with an * asterisk.

Preventive Measures
Air-intake filters prevent whiteflies from entering the garden. Isolate infected plants if possible.

Barriers & Physical Controls see *Controls: Section 1*

* Air-intake filters
  Aluminum foil
* Insect netting
* Vacuuming
  Water sprays

Pesticides see *Controls: Section 2*

Azadirachtin
Garlic
* Herbal oils
  Insecticidal soap
  Neem oil
* Pyrethrum
  Seaweed

Biological Controls see *Controls: Section 3*

Bacteria:
* *Saccharopolyspora spinosa* (spinosad)

Beetles:
  lady beetle
  (*Delphastus pusillus*)

Bugs:
* big-eyed bugs
  (*Geocoris punctipes*)
  damsel bugs
  (*Nabis alternates*)
* minute pirate bugs
  (*Orius spp.*)
  mirid bugs
  (*Marcolophus caligenosus*)

Fungi:
* *Beauvaria bassiana*
  Verticillium lecanii

Wasps, parasitic:
* *Encarsia formosa*

INTRODUCTION TO PART 2: CONTROLS — THE PROBLEM SOLVERS

You have read about the pests, with their voracious appetites, high reproduction rates, and evolutionary ability to adapt to adverse and stressful conditions. You also read detailed descriptions of the life and times of the pathogens, and the havoc they can wreak. After seeing their effects in your garden you might be in despair, or at least perturbed. You have reason to be; plant pests and diseases are not a laughing matter. Luckily, solutions are at hand.

Preventatives and controls specific to individual problems have been discussed in each chapter. The chapters conclude with a summary of controls. My preferred and favored solutions are noted with an asterisk (*). This selection is somewhat objective, but my subjective preferences did play a part in the selections.

This part of the book describes solutions in more depth. Many solutions can be used for more than one problem. This part of the book also describes choices of methods to resolve a problem.

Outdoors, control means elimination of noticeable damage to plants. Indoors, except in certain predator/prey situations, it means the elimination of the problem. Because of the speed of growth and reproductive prowess of herbivores, even a single pregnant female or a few spores can renew an infection that was almost eliminated.

This book was written with your health in mind, and the health of those who use the outcome of your efforts. The techniques, controls, pesticides, fungicides, and biological agents that are recommended are generally regarded as safe (GRAS).

There are no products with hard-to-pronounce ingredients or with names that are a compilation of chemical terms. Most of the solutions use nature-designed chemistry (mostly plant oils) or natural processes to protect the garden

This book shows you how to harness the power of many life kingdoms. Animals, fungi, plants, and other recently named kingdoms of organisms, as well as minerals and other benign substances, can thwart the efforts of herbivores of all kinds that would harvest the plant before you.

HOW TO USE PART 2: CONTROLS

First, if you haven't, read the Preventive Maintenance chapter, where you will find a wealth of information on how to protect your garden from the constant pest onslaught. Prevention is always the best policy, far easier than control.

Part 2: Controls is divided into four sections, each with effective strategies for a wide range of pests and diseases:

Section 1: Barriers & Physical Controls
Section 2: Pesticides & Fungicides
Section 3: Biological Controls
Section 4: Outdoor Strategies

Section 1 details methods of preventing the problem from getting to the plant or garden. Prevention is always the best policy.

Section 2 provides detailed discussions of the fungicides and pesticides mentioned in the first part of the book. All recommended solutions listed are safe for edible plants. In this section the active ingredients are listed in alphabetical order.

No matter what anyone tells you, and no matter how safe a salesperson or horticulturalist might say a fungicidal or pesticidal product is, if the label does not specifically say that it is meant for use with vegetables, it absolutely should not be used on marijuana in any stage of growth. A few popular naturally derived products that are not registered for use on edibles because of their danger to health are mentioned in this section to warn you not to use them. Ornamentals are not edibles—therefore products labeled for ornamentals must not be used on marijuana.

Synthetic pesticides and fungicides have been in use for fewer than a hundred years. They were first used in the 1940s and many have lost their effectiveness, as pests and pathogens have evolved genetic resistance. (See the Green Aphid Speech in the Aphids chpater.) At the same time, their use hurts environments and the beings in them, including humans and even pets. Although many new pesticides are now more targeted, they still continue to have unforeseen side effects. If you are interested in using synthetic chemistry, this controls section, in fact this whole book, is the wrong one for you because these substances are not discussed here.

Some of the recommended controls have been in development for more than 100 million years. Plants can't move away from predators like animals. Instead they have developed the art and science of chemical warfare to protect themselves from predators and diseases. This battle is ongoing. Just looking at a natural environment, you will notice that plants are quite successful in their efforts. Plants created the fragrances of herbs and spices as a warning to herbivores that their chemistry is toxic. You might have noticed that insects rarely bother fragrant plants such as cinnamon, garlic, lemongrass, onion, oregano, rosemary, sage, and thyme. That is plant chemistry at work. These plant substances are biodegradable and are not harmful to the environment as a whole—they are part of it.

Section 3 discusses biological controls. These are organisms that prey on our enemies but are benign to us. They are often referred to as "beneficials."

Herbivores—the cannabis eaters—and their predators and parasites have evolved together as well. Without predation the whole system would become imbalanced as the herbivores would eat and reproduce until scarcity levels were reached and the garden was consumed. In nature, the aphid lions and aphids will never lie together. Although this may be detrimental to particular pests, predators and parasites keep the pest population level lower so the lucky or more-fit survivors enjoy a well-stocked commissary. We use beneficials to eliminate the pest population.

If you decide to conduct research on your own you may find information on biological controls that are not mentioned here. Some organisms would be good candidates as controls but are not listed in the book because they are not commercially available. A few others are unsuitable for use in marijuana gardens because of their environmental requirements. There are also a few that have been omitted because of health concerns for people with compromised immune systems.

Section 4 covers some strategies that are particularly useful for the outdoor garden.

These four sections provide the gardener with a multitude of choices. There is usually more than

Many folk remedies can be found on the Internet. Only consider recipes using substances that are unlikely to cause harm to plants, pets, or humans. While there may be no harm in giving these remedies a try, the suggested solution should always pass this test:

1) Is it too far-fetched for you to trust?

2) Are there negative comments by others on the technique or substance?

3) Is this substance bad for pets, plants, or people?

4) Could this pollute if added to the water supply or runoff?

5) Will it attract any other hungry pests?

If any of the answers is "yes," avoid it.

one right solution, so picking one depends on your garden's particular situation and your style and comfort zone. Also, redundancy is a good technique. Attacking a pest using several controls is a surer, and often faster, method of stopping the threat to your plants.

The solutions in these sections are listed alphabetically by their generic names or ingredients. We also list a few brands. This is not an endorsement of these products specifically, but rather of the ingredients. Other brands with the same ingredients may be available and are just as useful.

Here's to a pest-free garden and healthy harvest.

SECTION 1.
Barriers & Physical Controls

Exclusion is the first defense against invaders whether they walk, fly, or crawl into your grow space. Barriers are designed to block pests from entry. Some invaders can be deterred by very simple means, saving you the trouble of messy treatments later. Many materials can be used to create barriers between plants and pests. Depending on the pest, the goal may be to block pests from reaching a specific destination: the plant canopy, the planting medium, or the entire garden. Suitable all-purpose barriers include horticultural shade cloth, paper, cardboard, and coir cover, which disrupt thrips' and fungus gnats' life cycle. A thick layer of perlite on top of the grow medium, or layers of mulch or gravel placed between the growing media and potential pests, also prevent or diminish an infestation.

Some barriers physically block pests.

Barriers prevent crawlers from climbing into the canopy from the ground or planting medium using sticky barriers

Herbs and spices repel pests using plant chemistry. Pests avoid some substances because they are toxic to that pest, smell bad, or possess some other quality that serves as a repellant.

Barriers can be quite effective, but they require diligent use. Any gaps in coverage or lack of thoroughness with a barrier method renders it useless. They often serve as good preventative measures.

Pesticides and fungicides, which are found in Controls: Section 2, control pests that have breached the garden or plant barrier with their eating habits, food supply, reproduction, or their respiratory or digestive functions.

SOLUTIONS
Air-Intake Filters (Indoors) For aphids, caterpillars, spider mites, thrips, and whiteflies.

Filter out airborne pests using filters on the air intakes. Place screens at either end of the ventila-

The Dust Shroom from Horti-Control helps keep aphids, thrips, and mites out of indoor gardens.

Phresh Filters makes carbon filters that serve as a barrier for indoor gardens, helping to keep pests at bay.

tion tubing to prevent infestation. Some insects, such as thrips, are so tiny that not all barriers of this type are effective. Make sure to use a screen designed to exclude thrips. To screen out the smallest insects indoors, cover all openings including vents, windows, louvers, and exhaust-fan openings. HEPA filters are the gold standard of filters because they screen out dust, pathogens, and the tiniest insects. There are several brands of screens designed for this purpose. They include Fly-Barr, BugBed 123, and No-Thrips. Even these will only screen out 70 to 80 percent, mostly the adults.

Aluminum Foil (Outdoors)

For aphids, leafhoppers, thrips, and whiteflies.

Foil or any reflective material placed on the ground around plants or over the growing media confuses flying insects including aphids, leafhoppers, and whiteflies, as well as thrips. Aluminum foil spread on the soil, beneath or around containers, is effective. You can also paint cardboard, paper, or plastic with silver paint. Aluminum-

coated heavy paper is also available commercially. These surfaces reflect UV and blue light upward, confusing the pests so they continue flying rather than landing to feast. Since the plants cover the reflective material as they grow, foil loses its effectiveness by mid-season unless there are wide spaces between the plants.

Baking Soda (Indoors) For ants and gray mold.

Baking soda (sodium bicarbonate) is a very alkaline substance that ants hesitate to step into, so you can make a line in the sand, or whatever your floor is composed of. It will protect the perimeter from incursions. Baking soda adds sodium to the soil, so be careful when exposing it to soil. See Section 2: Baking Soda for information about its use as a fungicide for controlling powdery mildew.

Bay Leaves (Indoors) For ants.

Bay leaves repel ants. Place the leaves along ant

trails and the creatures abandon them, or use the powdered leaves as a perimeter to protect an area.

Boric Acid (Indoors)
For ants, termites, and cockroaches.
See also Section 2: Boric Acid.

Boric acid is a stomach and contact poison effective for ants (see also the Ant Bait entry in Section 2 for more on boric acid's pesticidal uses). It has been used since the Roach Tablet product was introduced in 1922. Boric acid is a powder that sticks to ants' exoskeletons as they walk through a perimeter barrier. When the ants clean themselves, they lick it off, and ingest it. The acid activates as it moistens and it eats the ants' internal organs. Because of its toxicity to ants, it is effective as a food-based pesticide and as a barrier. When ants come in contact with enough of it, it clings to their exoskeletons and desiccates them. Boric acid is used medicinally by humans as an external application for skin conditions and is not considered hazardous to people or mammals. Boric acid adds the micronutrient boron when applied so care should be taken to prevent boron over fertilization.

Bottle Barriers (Outdoors)
For snails and slugs.

A 1- to 2-liter plastic bottle with the bottom cut out can be pushed into the soil about 2 inches deep around seedlings to keep out snails and slugs. Anchor the bottle in place with a stiff wire or coat hanger. Water the seedling through the hole in the top. Make sure there are no snails or slugs inside before you install the bottle.

Capsaicin (Indoors and Outdoors)
For ants, aphids, caterpillars, deer, gophers, grasshoppers, leafminers, moles, spider mites, thrips, and whiteflies.

Capsaicin is what makes hot peppers hot and pests don't like it. Plants produce capsaicin to deter herbivores. It is applied as a foliar spray. Capsaicin is more effective when combined with garlic (see Section 2: Garlic for more information). The downside to capsaicin is that sprayed leaves should be discarded, not used for edibles and other by-products.

The first product containing capsaicin was registered with the Environmental Protection Agency in 1962 to repel dogs and it is still available today. Hot pepper spray Bonide Hot Pepper Wax controls or repels ants, aphids, leaf miners, spider mites, thrips, and whiteflies. In addition to insects, capsaicin is used to repel larger outdoor pests such as deer, gophers, voles, moles, dogs, squirrels, and cats. BrowseBan is a safe capsaicin product used to repel deer.

RECIPE FOR CAPSAICIN PEPPER SPRAY

½ ounce (15 grams) dried or 4 ounces fresh medium-hot peppers (cayenne, Tabasco, or other hot pepper). Hot pepper sauce can be substituted. Use 1 to 2 tablespoons depending on strength.
¼ teaspoon lecithin granules
¼ teaspoon wetting agent
Enough water to make 1 pint of mix

NOTE: Wear gloves when handling the peppers and do not touch your mouth or eyes.

Grind the peppers and seeds in a blender. Add the water, lecithin, and wetting agent and mix. Strain the mixture through a tea strainer or cheesecloth into a glass jar. Use 1 to 2 tablespoons of the mixture to 1 pint of water and spray on plants. The mixture will become stronger as it sits, but it can also be used immediately. Test on a few leaves to make sure it doesn't harm the plant before spraying the whole crop.

Charcoal, Powdered (Indoors and Outdoors)
For damping off.

Powdered charcoal is used as a preventative for damping off. Dust seedling trays with powdered charcoal after planting. It is not necessary to apply more than once. The charcoal will not harm the seedling. However, powdered charcoal can be messy on the hands or clothes of anyone who comes into contact with it. Charcoal is used to improve soil tilth and provide support for soil microorganisms. It can be added at a rate of up to 5 percent of the soil by weight.

Cinnamon (Indoors and Outdoors)
For ants, caterpillars, crawling insects, damping off, fungus gnats, and spider mites.

Cinnamon can be used for several pests and diseases and in several forms. See also Section 2: Cinnamon; Ant Bait; Ant Brew.

Want to see ants go berserk? You probably have the means to do it right now, with cinnamon. Ants are totally repelled by it and it is an ant-icide. A barrier or slight coating of powdered cinnamon on the floor or on a perimeter barrier will keep ants from crossing the line. The powder remains effective for about two weeks. Cinnamon sticks last a bit longer. You can also make a paste of cinnamon powder and water and paint it onto the floor as a barrier and a repellent.

Cinnamon powder is also a preventative for

damping off. Dust seedling trays with cinnamon powder (or powdered charcoal, see above) after planting. Additional applications are not necessary. Cinnamon does not harm seedlings as long as the application is a "dusting" and not a heavy layer.

Cloves (Indoors)
For ants, caterpillars, crawling insects, and mites.

You would think that ants would want to spice up their lives. The fact is they hate cloves as much as cinnamon. Use either ground cloves or sprinkle unbroken cloves around. It works as well as cinnamon. See also Section 2 for a recipe for cinnamon/clove tea for mites.

Copper Stripping (Outdoors)
For slugs and snails.

A 2- to 4-inch-wide circular strip of copper reacts chemically with the slug or snail's slime and repels them with a small electric shock. Use a copper screen, flashing, or strips around the plant or the garden perimeter. Keep the surface of the copper clean for the best reactions. These bands are available commercially for use around the base of pots or around the base of single-stem plants. They are sold under a variety of commercial names but are generally known as "slug rings." Copper can also be used as an additive in pesticides. See Copper in Section 2.

Cream of Tartar (Indoors) For ants.

Cream of tartar repels ants and is found in nearly every grocery store. Trace an ant trail back to its entry point, whether it's under the door, under the window, or a crack in a wall or baseboard, and sprinkle liberally on and around the opening.

Diatomaceous Earth (Indoors and Outdoors)

For ants, fungus gnats, powdery mildew, slugs, snails, and thrips.

Diatomaceous earth is composed of tiny silicon crystals that were the shells of minute seafaring animals called diatoms. As these tiny sea creatures with hard silica shells died millions of years ago their shells sank to the ocean floor. They turned into deposits of chalky material that crumbles into a white powder.

The shells' sharp, glassy edges fatally injure and kill soft-bodied insects that attempt to navigate through them, such as fungus gnat larvae and thrips pupae. This substance also repels ants, slugs, and snails. The crystals are only effective when they are dry. They become soft when wet.

Pour a strip about 1-inch deep and an inch wide on top of the soil or floor. Slugs and snails also seem to have problems crawling over very dry surfaces and will avoid a thin layer of dry diatomaceous earth. However, it only works if the diatomaceous earth is kept very dry. Otherwise, the slugs will crawl across with no hesitation.

You can increase the effectiveness of diatomaceous earth by combining it with boric acid, cinnamon, or ground cloves. It can also be included in soil mix to increase the silica content.

Heat (Indoors) For aphids, damping off, fungi, mites, powdery mildew, thrips, and whiteflies.

Powdery mildew, some fungi, and several pests are heat sensitive. *Leveillula taurica* and *Sphaero-*

theca macularis, both forms of powdery mildew, stop growing at about 90 °F (32 °C) and die when the temperature rises to 100 °F (38 °C). However, raising the heat in an indoor growing area may not be a feasible option for every grower. First, it may be difficult to heat the space to such a high temperature. The second is that even a single peak of 100 °F (38 °C) affects the growth of plants. Vegetative plants may stretch a little from the experience, and buds in mid-stage growth, weeks three through seven, may also stretch a little. The heat treatment has relatively little effect on first and second week buds or on buds nearing maturity.

If you choose this method, minimize the heat's impact on the plants by heating the garden at the end of the day as the lights are turned off. During the dark phase, plants are not photosynthesizing and have lower water needs. They also don't have the stress of intense light as they open their pores to transpire water and cool down. The plants will have the evening to recover before the lights go back on.

If the plants are grown hydroponically, the use of heat can be counteracted by lowering the temperature of the water in the hydro system to 60 °F (15 °C). Keeping the roots cool will help the upper plant parts beat the heat. If you don't have a water chiller, add ice to the reservoir or flow-through system.

Roots of plants growing in soil can also be cooled using thermal ice packs at the base of the stem. The heat treatment should kill off most of the fungus and its spores. There may still be some fungal re-growth. These can be eliminated using spot treatments of a fungicide. Alternately, bottom heat can be used indoors to dispel some fungi, such as *Pythium*, which thrives at lower temperatures.

For greenhouse or indoor planting areas, using heat during a fallow period eliminates pests. Remove all plants, then heat the greenhouse or grow room until soil temperatures reach a minimum of 60 °F. Maintain that temperature for three weeks. Pest eggs in the soil hatch and the nymphs starve for lack of food.

Another method of using heat to exterminate many insects including thrips is to raise the air temperature to 104 °F and keep the humidity low for a minimum of one day. This method is only useful for empty or fallow spaces because these conditions are unsuitable for plants. Heat can also be used to sterilize media before use. See Section 4: Soil Solarization.

Horticultural Fabric (Outdoors)
For thrips, fungus gnats, and weeds.

Horticultural fabric is a weed barrier. The fabric is laid on the ground or on the container surface to prevent larvae from getting to the ground when they drop from the plant. The fabric is lightweight, lets water pass through easily, and can be cut with scissors, making it ideal for indoor use. Outdoors, it performs poorly because it tears easily, can't be walked on, and is carried by the wind unless it is secured around the entire perimeter. A better solution outdoors is using rugs turned upside down. They stay in position without being secured, can be walked on, don't tear, can be obtained for free from carpet stores, and last for years. In addition the rugs eliminate transpiration of water from the soil, and keep the soil cool on hot days. First lay down the rug, then cut holes using a box cutter where you want to locate the plants. Place drip irrigation heads under the rugs to prevent evaporation and to better direct the water to the root area.

Infected Plant Removal (Indoor and Outdoors) For all fungal diseases.

Let's say that one plant is infected with a few tiny white spots of powdery mildew on a few of its leaves. Remove it from the space before it infects other plants. Use a bag large enough to place over the diseased area, cover it, and then carefully cut or prune away the covered portion. Remove the bag from the room. This prevents the millions of waiting spores—the white powder on top of the leaves—from becoming airborne while being removed. Remember to wash your hands and clean the scissors or knife with soap and water, hydrogen peroxide, or alcohol.

Just as patients receive antibiotics after surgery to prevent infection, plants benefit from protection after surgical intervention, too. The area around the known infection should be sprayed with a fungicidal sealer or biocontrol agent such as *Trichoderma harzianum*.

The infection may reappear a few days after removing the leaves and spraying the plant. In this case, the plant may simply be susceptible to powdery mildew and as such, it presents a good location for the infection to start. It is much easier for the infection to spread to less susceptible plants once it has started. For this reason, the problem plant should be removed immediately by placing a bag over it and removing it from the space. Then spray the space with a fungicide once more. Many fungi attack new growth so make sure you protect the plant with a thorough spray.

Sometimes one plant attracts pests more than others and will serve as a primary host for other plants to be infected. Remove the plant before more damage occurs.

Insect Netting (Outdoors and Greenhouses) For aphids, caterpillars, fungus gnats, grasshoppers, leafhoppers, mites, thrips, whiteflies, and many other pests.

Insect netting presents a barrier that flying and crawling pests cannot penetrate. Place row covers over beds of small plants or drape the fabric over large ones. Reemay is one brand.

Screens are made of various mesh sizes to protect against different insects. The tightest weaves prevent adult thrips entry. Looser weaves are available for whitefly, aphid, and leafhopper protection.

Expect changes in light and temperature using insect netting. The screens reduce light by 10 to 15 percent and reduce airflow by 30 to 40 percent. This is an advantage in cool cloudy weather, but under sunny or warm conditions it can spell disaster without proper attention. First, rather than hanging low over the plant, where it acts like a blanket, holding a pocket of air, keep the curtain away from the canopy and sides of the plant so there is likely to be more air movement. If the temperature still has to be adjusted indoors you might use an inline fan with a mushroom cap or HEPA filter attached, or for even more cooling, a small swamp cooler.

Even the tightest thrips screen, with a hole size of 140 millimeters by 140 millimeters, allows about 15 percent of the pests—the juveniles—to slip through. It protects against all but the smallest mites, though, because they are considerably larger than thrips. A 300 millimeter by 300 millimeter hole screen protects against most aphids, leafhoppers and whiteflies.

To keep crawling insects out of the space bury the netting under the earth or attach it directly to the raised bed so crawling and walking insects are barred from entering. One convenient way to do this is to attach the bottom of the net to a piece of 1 inch by 2 inch board, wrapping it around several times, and then stapling it. Then bury the board beneath the soil line or attach it to the raised bed.

Moats (Indoors and Outdoors) For ants, slugs, snails, and all crawling pests.

Crawling insects must use pedestrian routes to get to their lunch. A moat presents an impenetrable barrier stopping them from getting to the plants. Indoors, table legs can be placed in trays filled with water so crawlers can't get from floor to table. Ants, snails, and slugs are stymied in their efforts trying to reach your plants. Moats can be used outdoors, too.

INSTRUCTIONS FOR A BEER MOAT

Pans filled with beer serve as a pit trap into which the slugs and snails fall and drown. Slugs and snails are attracted to the yeast in the beer. Dogs sometimes like to drink beer so the traps must be protected from canines. In rainy areas cover the traps so they don't flood.

Slugs and snails prefer their beer fresh, so the pans should be emptied and refilled every day or two if you are pouring beer. A homebrew is concocted by adding yeast to half a cup of sugar to a quart of water. Covered traps are available commercially but are expensive to install in a large area. Large coffee cans with an entry hole halfway down and a plastic lid on top work well despite beer-craving dogs and rain.

Mulch (Outdoors) For grasshoppers and root rot.

A thick layer of mulch prevents grasshopper nymphs from emerging in the spring. During the summer, mulch conserves moisture. Adding mulch or compost high in carbon to the top of the soil or mixing it into the soil at levels up to 20 percent decreases the viability of *R. solani*, a pathogen causing root rot.

Pest Removal (Indoors and Outdoors) For caterpillars, mealybugs, slugs, and snails.

Pest removal by hand is often the fastest way to get an infection under control. Caterpillars, mealybugs, slugs, and snails are large enough to see, and move slowly so they are easy to capture. Remove them with a cotton swab dipped in alcohol, or just a cloth. Mealybugs tend to locate in plant crevices and other hard-to-get-to spots. Q-tips moistened with isopropyl alcohol are ideal tools. Remove and crush individuals from the crevices of the plants with a cloth or sponge.

Go slug and snail hunting at night when they emerge from their hiding places or on rainy days when snails and slugs are most active. Use a flashlight to find them. Squash them in place, or place them in a can half-filled with water. Plastic gloves are advisable. When the hunt is over, feed them to chickens, add them in the compost pile, or leave them to decompose naturally, and for predatory insects.

Potassium Bicarbonate (Indoors and Outdoors) For crawling insects and fungal diseases including powdery mildew.

Potassium bicarbonate creates a very alkaline surface that crawling insects avoid. Fungi germinate and grow best in acid environments. Potassium bicarbonate creates an alkaline environment that prevents fungi and mold germination and stymies growth. Potassium is a fertilizer that the plant eventually absorbs through its stomata or through its roots if it is washed off.

Repotting Substitute (Indoors) For ants and root aphids.

For most plants, repotting by removing soil from the roots is a stressful experience that sets them back weeks. With ants or root aphids, if a single plant is infected and is not valuable, it might be easier to just discard it. Then you are removing sources of infection before it reaches the rest of your plants.

So what do you do when you have a plant in soil with pests and you don't want to repot? I've had to deal with that problem. A mother plant was growing in a 15-gallon container and was more than 5 feet tall, with ants in the soil. My solution: First I placed the container in a plastic drip tray and added enough water to cover the drainage holes. That way the ants couldn't escape from the bottom of the container. Then I sprinkled powdered cinnamon on the top of the container. This prevents ants from escaping at the top because of their aversion to cinnamon. Then mix a brew of 5 ounces of 1 percent Orange TKO (a natural cleaning product made from D-limonene) and 5 ounces powdered cinnamon in 2 ½ gallons of warm water. Let it cool and pour the mixture into the plant. Ant workers will come up through the soil, evacuating with eggs, but they are dying. The rest of the colony dies in the nest.

Another method is to use pyrethrum or insecticidal soap in water solution, and water the container thoroughly.

Screening. See Insect Netting.

Sticky Barriers (Indoors and Outdoors)
For ants, aphids, caterpillars, fungus gnats, leafhoppers, thrips, and whiteflies.

Many insects are loathe to venture onto sticky surfaces since that situation can easily be the death knell if the insect can't break free. For insects that don't fly, sticky surfaces may completely block access to a plant. For aphid-herding ants, this may also cut off the path to their herd, making the aphids much more vulnerable to the environment and predators.

Gardeners should never put a sticky substance directly on the stem. This may affect the plant adversely. Instead, wrap paper, aluminum foil, or plastic around the stem and paint that with a sticky substance to prevent any creepers from climbing up. Other options: Use old toilet paper or paper towel tubes to protect the bases of plants. Cut them lengthwise, wrap, and staple, taking care not to damage the plant. Wrap masking tape sticky side out. Tanglefoot or Stickem are commercial sticky adhesives.

Sticky Cards (Indoors) For aphids, fungus gnats, leafhoppers, thrips, and whiteflies.

Sticky cards are an excellent way to monitor the garden for pest infestations but a poor method of control. They are effective for a number of pests. These rectangular cards are bright yellow in color, which attracts aphids, fungus gnats, leafhoppers, and whiteflies. Blue or bright pink cards are used to attract thrips. The cards are coated with an incredibly sticky glue. For monitoring purposes, hang one card for every 20 square feet of growing space.

It is often more convenient to hang the yellow cards vertically 6 to 12 inches above the top of the canopy. Shake the plants periodically to disturb any insects present and encourage their ensnarement.

To detect thrips, position the cards so the bottom of the pink or blue card hangs about 2 inches above the plants. The number of thrips trapped is an indicator of the density of the infection. Lurem-TR (Koppert Biological) attractant lures two to three times more thrips to the cards. Remove sticky cards before releasing biological controls as they, too, will be attracted to the cards. Outdoors, petunias can be used as indicator plants. See Section 4: Plants as Indicators.

Homemade yellow sticky cards can be made by painting thin rectangles of ¼-inch plywood or paper with yellow paint and smearing them with Tanglefoot or Stickem, or a homemade substitute of petroleum jelly or vegetable oil. The homemade sticky concoctions have the advantage of being easy to clean off with a little soap and water so the boards may be reused. Cards placed beneath benches and around air-return areas can provide additional information on emerging populations. Colored cards are not an adequate control when used alone. They are best used as simple indicators to help identify pests in the garden.

Talcum Powder (Indoors and Outdoors)
For ants.

Talcum powder is an ant repellant, and can be used to create a barrier by encircling the area to be protected. Chalk and baby powder are two commonly available sources of talcum. Sprinkle powder wherever you see a trail of ants and they will avoid this area next time. Ants don't seem to like talcum powder much, especially baby powder.

Vacuuming (Indoors and Outdoors)
For aphids, caterpillars, fungus gnats, spider mites, thrips, and whiteflies.

Vacuuming insects off plants with a shop vacuum helps control the population but is unlikely to eradicate it. Vacuuming can also be used to lower the pest population before using other means of destroying it. Lowering the population means that there is an immediate cessation of the stress on the plant caused by thousands of insects or mites sipping from the fountain. Vacuuming makes it is easier to completely exterminate pests using other methods.

In the cool of the night and early in the morning, insects such as fungus gnats, thrips, and whiteflies are slower moving and easier to vacuum up. Stir the foliage and capture any that fly up.

Vinegar (Indoors and Outdoors) For ants.
Vinegar is an ant repellant and can be used to break up an existing ant path. It acts as a deterrent, and interferes with the pheromone scent trail.

Water Spray (Indoors and Outdoors)
For aphids, caterpillars, mealybugs, spider mites, and whiteflies.

Hose off infested plants regularly with a moderately strong stream of water. The force of water will wash insects off of the plants.

Outdoors, where the plants dry quickly in the warm sun after washing, hosing down the plants knocks off insects and mites and dislodges eggs, reducing the population. This reduces stress on the plant before introducing beneficials or attempting other controls. Misting is a favored spider mite treatment, as it also raises the humidity.

SECTION 2.
Pesticides & Fungicides

Pesticides and fungicides are designed to control or eliminate unwanted insects or fungi that are detrimental to plant life, while leaving all other forms of life unharmed. This is usually accomplished by targeting a vital function within an insect or fungus and disrupting it. There are many strategies, but most often it involves interfering with food or reproduction. Pesticides may cause a specific pest to find their usual food source repellant, or may turn a food source into a toxin for that particular insect. Other strategies interfere with the ability of pests to mature into adulthood, or may render them unable to reproduce. Finally, others interfere with bodily systems such as respiration or digestion, or might even cause confusion, leaving pests vulnerable to predators.

There are many commercial pesticides and fungicides available but it is critical that the gardener use only solutions that are designated for edibles since cannabis may be smoked, vaporized, or eaten. If you wouldn't put it on a tomato and eat it, then you shouldn't put it on your cannabis plants. Any product labeled only for "ornamental use" means that it is not intended to be used on food crops. Cannabis is often grown and used by individuals with health issues or compromised immune systems, so plants should be grown using only the safest pest control products.

In addition to commercial plant protectants, you can make your own effective, nontoxic, low-impact insecticides using household ingredients.

Rmember: Never buy commercial or garden-store treatments to treat cannabis that are labeled "for ornamental use only." This means the product is not intended for use on plants that will be consumed.

This section covers active ingredients listed in alphabetical order. A few products are referenced, but there are other brands available. Once you have chosen a solution, look for products with these ingredients.

SOLUTIONS
Alcohol Spray (Indoors and Outdoors)
For mealybugs and scales.

A 50/50 dilution of 90 percent isopropyl alcohol and water is an effective method of killing mealybugs. Spray or dab directly on the bugs once every other week four times. This disrupts the life cycle.

Aluminum Silicate. See Silica.

Ant Bait (Indoors and Outdoors) For ants.

Ant baits are products that contain both a poisonous substance and a compound irresistible to ants. These products are often boric acid based. The idea is to get the ants to carry the bait back to their nest where it kills both workers and queens.

Some products use boric acid and are especially popular as ant bait and are very safe. Other active ingredients include biologically based toxins spinosad or avermectin. Registered ingredients sulfluramid or fenoxycarb are used in others.

Ant baits and ant stakes are very safe methods of dealing with the problem because they use minute amounts of poison in a very targeted way. The ingredients are not released into the general environment. Instead, the target acquires the slow-acting poisons. The affected ants have time to interact with a large number of nest mates so the poison affects a large group.

Examples of these products are Drax Gel Ant Bait, Borid, and Dr. Moss Liquid Ant Bait. Other commercial brands available include: Green Light Fire Ant Killer with Spinosad, Valent Esteem Ant Bait, Clinch Ant Bait, Spectracide, and Combat Outdoor Ant Killing Stakes.

You can make your own ant bait.

RECIPE FOR SUGAR-LOVING ANTS:

Mix equal parts sugar or corn syrup and boric acid or Borax detergent and place small amounts in shallow dishes near the infested area.

Recipe for both sugar-loving and protein/grease-loving ants:

4 tablespoons peanut butter
6 tablespoons honey
¾ teaspoon boric acid

Mix the ingredients and put in shallow containers near plants. Outdoors, the containers must be protected against predation by dogs, raccoons, and other wildlife.

Another version:
1 package gelatin (makes about 1 cup gelatin—flavored is okay)
1 tablespoon honey
1 tablespoon boric acid
1 egg

Prepare gelatin per directions for the dessert. As the gelatin cools, add a tablespoon of honey and boric acid. Once it is cool enough that it will not immediately cook the egg, stir in the raw egg. Let gelatin set in small cups. Once they are set, place the cups in areas that are accessible to the ants. Store extra cups in the freezer.

Ant Brew (Indoors) For ants.

First mix the ant brew by mixing together these ingredients:

1 ounce 1 percent orange oil such as TKO (a natural cleaning product made from D-limonene)
1 ounce ground cinnamon
1 gallon warm water

Sprinkle ground cinnamon lightly on the top of the medium in the plant container. Ants are repelled by cinnamon, so now they won't escape through the top. Water thoroughly with the ant brew until it drains out

Avid (Indoors and Outdoors)

Do not be misled to use this product on marijuana. This product and others composed of abamectin are approved only for ornamentals, not edibles or smokeables. It is a nerve poison and affects the reproductive system.

Azadirachtin (Indoors and Outdoors)
For aphids, leafhoppers, mealybugs, scales, root aphids, whiteflies, and other sucking insects.

This compound is made when a limonoid is processed from neem oil using alcohol. It is a growth and reproductive inhibitor.

It is relatively harmless to butterflies, bees, and other pollen-eating insects, and to beneficial predators and parasites because it must be ingested to be effective. It breaks down in five days in the presence of light and water.

Once ingested, the insect is unable to feed, molt, or lay eggs and soon dies. For use in hydroponic systems, mix as per label directions and add to the reservoir. AzaSol is a water-soluble form of azadirachtin. Other azadirachtin products are marketed commercially as Azatrol, Bioneem, and AzaMax, among others. Azatrol may turn the water a milky white at first or produce foam or froth. This is a normal reaction.

See also Section 2: Neem Oil.

Baking Soda (Indoors and Outdoors)
For ants, powdery mildew, and gray mold.

Sodium bicarbonate is baking soda, found in kitchens. It controls powdery mildew because it raises the pH of leaf surfaces. Rose growers have used sodium bicarbonate for powdery mildew for more than 70 years. Potassium bicarbonate (in this section) works the same way, but as its close cousin, baking soda is often preferred by gardeners because it is so readily obtainable. However it is not as effective as potassium bi-

carbonate and leaves sodium in the soil when it breaks down.

RECIPE FOR BAKING SODA SOLUTION
1 teaspoon baking soda
Few drops of castile soap or other wetting agent
1 pint water

Mix ingredients together in a spray bottle. Apply the spray weekly on new growth.

Black Pepper Spray (Indoors and Outdoors)
For caterpillars and other crawling insects.

This is used both to prevent predation by caterpillars and to kill them. Recipe:
1 tablespoon ground black pepper
1 quart water
3 ounces potable alcohol such as 100-proof rum or vodka (optional)
Few drops castile soap or wetting agent

Black pepper contains irritants that disrupt caterpillar feeding and affect the skin. The alcohol dissolves toxins not soluble in water. Mix and let stand for a day. Spray plants.

Bordeaux Mixture (Indoors and Outdoors)
For powdery mildew, gray mold, and other fungi.

Bordeaux mixture is a combination of copper sulfate and hydrated lime that was first used in European vineyards for the control of diseases such as downy mildew. In the early to mid-1800s, European grapevines, *Vitis vinifera*, were struck by a succession of diseases and pests to which they had no resistance, introduced by grapevines collected from North America. Downy mildew, in particular, ravaged the vineyards. While searching for a solution, botany professor Pierre-Marie-Alexis Millardet of the University of Bordeaux noticed that grapevines near roads did not seem to contract the disease as readily as vines further into vineyards.

After investigating, he found that these vines had been sprayed with a combination of copper sulfate and hydrated lime to make the grapes undesirable in appearance and taste to passersbys, so they wouldn't stop and pluck a few grapes to eat. He conducted further tests with the mixture and published his results in 1885.

Bordeaux mixture works because the copper ions affect enzymes in the fungi and prevent the germination of spores. It is best used as a preventative, before disease has started. It controls both downy mildew and powdery mildew. Bordeaux mixture is available commercially in a premixed package. Two types are Dexol Bordeaux Fungicide or Hi-Yield Bordeaux Mix.

The down side is the copper leaches into rivers and streams and negatively affects livestock, fish, and earthworms. Except in cases of rampant infection, Bordeaux mixture is not recommended outdoors because of these risks. Try other fungicides first for smaller outbreaks.

Boric Acid (Indoors and Outdoors)
For ants, cockroaches, and termites.
Boric acid is a stomach and contact poison. (Hence its cross listing in Barriers and Physical Controls. See Section 1: Boric Acid for more details on its use as a contact poison.) As a food-based pesticide, boric acid works once it is wetted by the ant's digestive system. It liquefies the ant's internal organs. Boric acid is safe for bees and birds, and has medical uses for humans as an external application for skin conditions. It is considered non-hazardous to other mammals, and is okay for use on edible plants.

Bt. See Section 3: *Bacillus thuringiensis*

Capsaicin (Indoors and Outdoors)
For ants, aphids, caterpillars, deer, gophers, grasshoppers, moles, spider mites, and thrips.
Capsaicin is what makes hot peppers hot and pests don't like it. Plants produce capsaicin to deter herbivores. It is applied as a foliar spray. Capsaicin is more effective when combined with garlic. See Section 1 for products and a spray recipe.

Chamomile Tea (Indoors and Outdoors)
For gray mold and damping off.
Chamomile blossoms contain sulfur, which suppresses damping off. Make a tea using ¼ cup of chamomile blossoms by pouring a pint of boiling water over the blossoms and allowing the tea to steep until cool. Strain this liquid into a spray bottle and mist seedling trays before planting. This tea can also be used anytime after the seeds have been planted. Chamomile tea can be used as a seed soak immediately before planting. Chamomile blossoms and teabags are available at health food stores and organic grocery stores.

CINNAMON
Cinnamaldehyde (Indoors and Outdoors)
For ants, aphids, gray mold, fungus gnats, powdery mildew, spider mites, thrips, and whiteflies.

For several years the products Cinnamite and Cinnacure, composed of cinnamaldehyde, were available and widely used in agriculture and by marijuana growers. However, currently these products are difficult to find.

Cinnamon Oil and Tea (Indoors and Outdoors) For most crawling insects, damping off, gray mold, mites, powdery mildew, and thrips.

Cinnamon tea controls powdery mildew, providing effective control, but not eradication. Cinnamon also enhances other sprays and can be used in combination with other herbal sprays such as clove and coriander. It is very effective against insects, both as a repellant and insecticide.

RECIPE 1 FOR CINNAMON TEA
Mix food-grade cinnamon oil to a ratio of one part oil to 200 parts water, or 1 teaspoon to a quart of water. Rosemary oil and thyme oil can also be added to the cinnamon oil. The total concentration of oils in the solution should never be more than 0.75 percent.

RECIPE 2 FOR CINNAMON TEA
Boil 1½ pints of water, remove from heat, and add half an ounce of powdered cinnamon. Let the tea cool to room temperature and add ½ pint of 100-proof grain alcohol and let it sit for a day. Strain out the particles of cinnamon and use as a spray.

CINNAMON-CLOVE TEA FOR MITES
Brew this combination tea using powdered cinnamon and cloves, available at the grocery store. Use 1 ounce of each to a gallon of water. Bring the water to a boil, then remove from the heat, and add the cinnamon and cloves. Let the mixture stand until the water has cooled. Strain and use as a foliar spray. Substituting a quart of 100-proof grain neutral spirits, vodka, or rum for a quart of the water draws out more water-soluble oils, making the spray more effective.

Citrus Oil. See D-limonene.

Compost Tea (Indoors and Outdoors) For damping off, gray mold, Verticillium wilt, and root, leaf, fungal, and bacterial diseases.

Compost tea is used to introduce beneficial organisms, provide nutrients that protect plants from pathogens, and increase vigor and growth. The nutrients are readily available and the organisms work in various relationships with the roots. The tea also produces enzymes and hormones that stimulate plants.

RECIPE FOR TRADITIONAL COMPOST TEA
Traditional compost tea is made from mature compost, a beneficial addition to the garden in any form. Make this tea by filling a net bag, stocking, or paint strainer with compost and dangling it in a container filled with unchlorinated water. Use one part compost by weight for four parts water. Sometimes additional nutrients such as unsulphured molasses are added to the mix to increase the microorganisms. Use about 1 to 2 tablespoons per pound of compost.

To maintain aerobic conditions the water should be vigorously circulated and well-aerated during the entire process. Place a strip bubbler attached to an aquarium pump to the bottom of the container to circulate the water. Steep the compost for two to three days, then turn off the

The Bountea Garden Tea Brew Kit is a home brewing option that simplifies making compost tea, helping to provide nutrients and microbes for maximum plant health.

pump and let the mixture settle. The tea should be dark brown and have no sour smell. If the tea smells of ammonia or smells rotten, this means the tea has become anaerobic and should not be used.

You can use worm castings—vermipost—rather than compost. It is much higher in nutrients than compost so use about 4 ounces per gallon of water, a little more than a pound in a 5-gallon bucket. Add 1 to 2 tablespoons unsulphured molasses per pound of castings. Follow directions in previous paragraph.

> Unchlorinated water must be used to brew compost tea because the chlorine kills the very microorganisms we hope to culture. You can use rain or pond water. If you are using city water, here are several ways you can season it. It can sit in the sun with an aerator going for about an hour, or you can add 1 tablespoon ascorbic acid (vitamin C) per 75 gallons of water.

Another kind of compost tea is made by culturing micro-organisms in aerated water enriched with nutrients such as composted forest humus, chicken litter, worm castings, guano, seaweed, fish emulsion fertilizer, gypsum, and oyster shell. Kits are available that supply the aeration system, the microorganisms, and the nutrients. Some hydroponic stores sell the compost tea ready-to-use. This kind of tea has several advantages over traditional teas because there is more control of the ingredients and the microorganisms inoculated into the brew.

Compost tea should be used fresh, as soon as possible. Strain the tea through a pillowcase, metal screen, cheesecloth, or lace curtain material. Use it as a foliar spray. This helps protect the leaves from pathogens and provides nutrients. It should also be used as a soil drench or in a drip-to-drain hydroponic system.

Copper (Indoors and Outdoors) For damping off, fungi, gray mold, and powdery mildew.
Copper has a long history of use as a fungicide. It is highly effective against powdery mildew, gray mold, and *Pythium*, but can injure plant tissue. Formulas use various compounds, which differ in phytotoxicity. Don't apply copper fungicides in cold wet weather, when copper ion availability increases and its toxicity increases.

When copper is in a water solution, its ions pass through the fungi's cell walls and attack proteins in the cells, killing them.

Take the same precautions using these sprays as you would with Bordeaux mix. Outdoors, use it only before flowering. The spray can be more controlled indoors, so you can use it on leaf areas, but not on the bud. Copper fungicides are implicated in the toxicological problems of farm workers who are in constant contact with them. However, copper has been used in home gardens and orchards for centuries with few problems.

Ingestion may have health consequences so precautions should be taken. Use protective clothing. After harvest throw sprayed leaves away—don't use them for cooking or kiefing.

Use this spray as a last resort when you are dealing with a heavy, out-of-control infection. Outdoors, where spray application is at the mercy of prevailing breezes, copper should only be applied before flowering.

There are many ready-to-use copper fungicides available. Some of them are Copper-Count-N, Concern Copper Soap Fungicide, and TopCop.

Coriander Oil (Indoors and Outdoors)
For apids, caterpillars, fungi, fungus gnats, mealybugs, scales, molds, powdery mildew, and spider mites.

Coriander oil is extracted from coriander seeds. It inhibits fungal infections in cases of certain anaerobic fungi. Add it directly to plant nutrient solutions or apply it as a foliar spray to treat molds, mildews, insects, and mites. It is nontoxic to humans, animals, and plants and is biodegradable and environmentally friendly. One product, SM-90, contains coriander oil, canola oil, and a pH buffer.

D-limonene (Indoors and Outdoors)
For *Pythium, Fusarium, Rhizoctonia*, and other molds and fungi, ants, aphids, fungus gnats, mites, root aphids, scales, thrips, and whiteflies.

D-limonene is one of the essential oils from the rinds of citrus fruits including oranges, lemons, and limes. It smells like oranges. It is the active ingredient in citrus-scented cleaning products.

D-limonene's mode of action is similar to pyrethrum. It both repels and paralyzes insects. It is available in a variety of formulations to kill and repel ants, hornets, wasps, fleas, and other insects. It is harmful to beneficial insects if sprayed directly on them. Commercial products include Ortho Home Defense Indoor Insect Killer, Clean Green All Purpose Cleaner, TKO, D'Bug Safer Brand Ant and Roach Killer, Sure-fire Crawling Insect Killer, and Orange Guard Fire Ant Control.

Ferric phosphate (Outdoors)
For slugs and snails.

Ferric phosphate ($FePO_4$) is a natural product that is ubiquitous in nature. It comes coated with a bait attractive to slugs and snails. When they ingest it, the iron phosphate interferes with calcium metabolism in their gut. Once ingested, they stop eating immediately and die in five days. Sprinkle it in a perimeter pattern over a larger area or across snail/slug paths. It is safe for humans and pets. Brand names include Sluggo and Escar-Go!

Garlic (Indoors and Outdoors)
For ants, aphids, caterpillars, damping off, deer, grasshoppers, gray mold, molds, powdery mildew, root rot, rots, spider mites, thrips, and whiteflies.

Garlic (*Allium sativum*) is a member of the onion family and is good for repelling vampires and many other pests. It has antifungal and antibacterial qualities as well. Garlic can be used as a prophylactic, before there is a sign of infection, or as a method for addressing specific pests. It is more effective when used preventatively rather than for battling a severe infection. Apply about every 10 days. The odor lingers so it should not be used for two weeks before harvesting. Garlic spray is rinsed from the leaves by rain so it should be reapplied when the storm has passed.

Garlic has several pathways for destroying fungi, including its high sulfur content, which creates an inhospitable environment for fungal spores to germinate. It is a general-purpose insecticide, but must be used with caution indoors and outdoors because it also kills beneficial insects. Don't use it when beneficial insects are in use. It can be used as a spray on its own, or used in other sprays as an additive, and also used as a preventative. Grasshoppers are attracted to garlic mixtures, so it can be used on trap plants in a bait mixed with boric acid. The grasshoppers eat the plants with the garlic mix and are slowly killed by the boric acid.

GARLIC SPRAY

Make a spray from ground garlic, sesame oil, and a wetting agent or castile soap.

1 quart warm water
2 ounces sesame oil
8 ounces ground garlic
Wetting agent as directed or ¼ teaspoon
*   castile soap*
1 teaspoon lecithin

Mix the garlic and sesame oil and let stand 24 hours. Add water, lecithin, and soap and store with tight cover in refrigerator. Use ½ cup in 1 quart water. Spray directly on plants.

RECIPE ENHANCEMENTS

Add

2 ounces 100 proof or higher drinking alcohol
*   such as rum or vodka per quart of mix. The*
*   pesticidal oils dissolve in the alcohol.*

Make a combination pesticide by adding capsaicin pepper with the garlic. It is even more effective.

RECIPE 2 INSTANT GARLIC SPRAY

1 tablespoon garlic juice
1 quart water
Wetting agent as directed or ¼ teaspoon liquid
*   castile soap*
½ teaspoon lecithin

Combine and spray directly as needed.

Garlic-based pesticides include Dr. Earth Pro-Active, Garlic Barrier, and Envirepel for ants, aphids, caterpillars, spider mites, thrips, and whiteflies. Garlic fungicides for powdery mildew include Citrall Lawn and Garden Fungicide.

Greensand (Indoors)

For ants, aphids, caterpillars, and many other pests. See Silica, this section.

Harpin (Indoors and Outdoors)

Pest and disease resistance.

Harpin is a naturally occurring protein that is the active ingredient in a new type of pesticide. Supposedly it stimulates the plant's natural defense system, and promotes more vigorous growth and hardier plants that are more resistant to disease with increased yields. The protein was discovered in the pathogenic plant bacterium, *Erwinia amylovora*, which causes fire blight in pears and apples. Employ is the only product available with this technology, the efficacy of which is still being explored.

Herbal Oils (Indoors and Outdoors)
For ants, aphids, caterpillars, fungus gnats, gray mold, mealybugs, powdery mildew, scales, spider mites, thrips, and whiteflies.

Herbal oils are made from a range of plants, including cinnamon, citronella, clove, coriander, eucalyptus, lavender, lemongrass, mint, oregano, rosemary, and thyme. They can be used closer to harvest because they will evaporate in five to 10 days and do not leave a residue.

These oils are Mother Nature's repellents, and contact insecticides that deteriorate the bodies of spores, insects, and mites. When developing my herbal oil products, oils from plants that were not bothered by mites, ants, aphids, and powdery mildew were selected. By blending together several oils, there is a multi-prong attack: Some oils deteriorate the exoskeleton of the pest, while others destroy the internal tissue, and others the nervous system.

Multiple applications of herbal oils will break up the life cycle of the pest. With spider mites, spraying three times, three to four days apart, will eliminate them by knocking out hatchlings.

Mist plants late in the day or late in the light cycle for indoor gardens. Always spray directly on pests that often reside on the underside of leaves. Store herbal oils in a cool, dark place with the cap firmly sealed.

Availability of commercial pesticides and fungicides based on herbal oils has increased. Dr. DoRight's blends oils as a defense against powdery mildew. Miticides include SNS-217 Spider

Mite Control and Rasta Bob's Death Mite. Ed Rosenthal's Zero Tolerance will eliminate mites as well as a wide range of soft-bodied insects and mold diseases. If the pests are in the soil, use a soil drench product like SNS-203.

Foggers

Foggers emit a fine spray that spreads over the garden. An herbal oil or pesticide spray repels insects and exterminates pests already in the space. The fog lays down a thin layer of herbal oil or pesticide on the surface of the leaves, also preventing colonization by fungi. Herbal oil pesticides that contain cinnamon, rosemary, clove, mint, other herbs and spices, or pyrethrum, can be used. Ed Rosenthal's Zero Tolerance and other commercial herbal sprays can be used in garden foggers. Dilute the ready-to-use products at a rate of up to five to one. Make sure that the plants aren't adversely affected by testing the spray on a few branches before general use.

Horticultural Oils (Indoors and Outdoors)
For aphids, fungus gnats, mealybugs, scales, spider mites, thrips, and whiteflies.

These light oils smother insects by clogging their spiracles, or breathing holes. Insects must be directly exposed or sprayed for these solutions to work. Some oils are derived from vegetables while others are petroleum-based. Plant-based oils deteriorate over time, but will leave a residue on the leaf surface. Petroleum products should not be used on the buds or on leaves intended for cooking or kiefing.

Horticultural oils are either winter, or "dormant" oils, or summer oils. Summer oils are for use on actively growing plants. They are light and highly refined. Winter oils, or dormant oils, are heavier and are meant for outdoor use over the winter. Winter oils severely damage growing tissue, so avoid them and use only summer oils. Be sure to check the label of any product purchased to make sure it is the right type of oil for your garden.

Growers use oil sprays to prevent and cure fungal infections. Until recently most horticultural oil sprays were made from petroleum distillates. However, organic growers have switched to using botanical oils. Aside from the safety factor, botanicals such as cottonseed, jojoba, and sesame oils have fungicidal properties. They can be used in combination with other spray ingredients listed here. The oils are mixed at 1 to 2 percent concentrations. A 1 percent solution is about a teaspoon per pint, 3 tablespoons per gallon, or half a gallon in 50 gallons. Add a wetting agent to help the ingredients mix. Horticulture oil sprays should only be used on the leaves, not the buds. Use weekly on new growth. The oils have no residual effects and must be reapplied regularly.

Commercial products include Green Light Horticultural Oil, Dr. Earth Pro-Active, and GC-Mite

Hydrogen Peroxide (Indoors and Outdoors) For ants, aphids, all fungal infections, fungus gnats, and spider mites.

Hydrogen peroxide (O_2) is the first-aid treatment in the brown bottle in the pharmacy or grocery store. It is a contact disinfectant that leaves no residue and dissipates into water and oxygen as it evaporates. O_2 has no residual effect. It is a great prophylactic treatment because it kills any spores and organisms with which it comes in contact, whether on table and wall surfaces, or on the plants themselves. It is used to sterilize equipment and spaces and can be used in combination with other treatments.

Hydrogen peroxide sold in the pharmacy is diluted to 3 percent. Garden shops sell 10 percent O_2. OxiDate uses a peroxygen formulation that is a very active form of hydrogen peroxide.

Use 3 percent O_2 full strength to sterilize tools, equipment, and surfaces. Mix it one part to two parts water to make a 1 percent solution that can be used on plants, about 5 ounces to 11 ounces of water to make a pint-of-water treatment, or a quart of O_2 in 2 quarts of water. When using horticultural-grade 10 percent, use 3 tablespoons per pint, or 10 ounces with water to make a gallon.

Insecticidal Soap (Indoors and Outdoors) For ants, aphids, grasshoppers, leafhoppers, mealybugs, spider mites, scales, thrips, and whiteflies.

Insecticidal soaps are potassium or ammonium salts derived from fatty acids. They damage the exoskeletons of soft-bodied insects and cause them to dry out, and gum up the spiracles found along the body, through which arthropods exchange CO_2 for oxygen. Insects must be sprayed directly with a thorough coating to kill them. Once the soap dries, it is no longer lethal. Apply weekly as a preventative. These products have little effectiveness against eggs and do not harm beneficial insects that are not caught in the spraystorm. The soap is safe to use up to the day before harvest.

To make your own insecticidal soap use water with a neutral pH that is not so "hard" that it develops a scum at the surface when soap is added. The spray is more effective when the soap dries slowly, so spray at dawn or dusk. Adding vegetable glycerin or vegetable oil to the recipe extends drying time.

RECIPE 1

1 tablespoon liquid castile soap such as Dr. Bronner's peppermint or pure castile
1 quart water

Add soap to water. Spray.

RECIPE 2

1 tablespoon liquid castile soap
1 quart water
2 tablespoons vegetable glycerin or 1 tablespoon sesame oil.

Use ½ teaspoon lecithin granules if you use sesame oil

ENHANCEMENTS:

1 teaspoon garlic juice
½ ounce capsaicin liquid such as hot sauce
¼ teaspoon cinnamon or rosemary oil
Mix in bottle and spray.

Ready-made brands include Safer Brand Insect Killing Soap, M-Pede, Neodorff's Insecticidal Soap, and Monterey Quick. Always test soap solution on a few branches before spraying the whole crop.

Limonene. See D-limonene.

Manure Tea (Outdoors)
For ants.
Drench anthills with manure tea or compost tea. Repeat treatment daily until problem is resolved.

Milk (Indoors and Outdoors)
For powdery mildew and gray mold.
Milk kills powdery mildew—that discovery made rose growers very happy. Scientists are not exactly sure just what milk does, but it does seem to work very well as a preventive and control measure.

MILK SPRAY FOR POWDERY MILDEW

Use one part milk to nine parts water. Whole, 2 percent, 1 percent, or skim milk can be used. Recipes with a ratio of up to one part milk to five parts water are acceptable. Using more than 30 percent milk results in the growth of a benign mold on the top of the leaves.

PREFERRED ENHANCEMENT:

Per quart of milk-water mix, add 1 teaspoon potassium or sodium bicarbonate. These act to prevent new growth of fungi such as powdery mildew.

OTHER ENHANCEMENTS:

1 teaspoon garlic juice
½ ounce capsaicin liquid such as hot sauce
¼ teaspoon cinnamon or rosemary oil

Mycorrhiza. See Section 3: Mycorrhiza.

Neem Oil (Indoors and Outdoors) For ants, aphids, fungus gnats, grasshoppers, gray mold, leafhoppers, mealybugs, powdery mildew, root aphids, scales, spider mites, thrips, and whiteflies.
In India and Asian countries, neem tree products are used for many medical, personal hygiene, and beauty products. Neem is also an all-purpose natural pesticide. Aside from arthropods, it also

provides protection from nematodes, and fungal and bacterial diseases.

Neem oil contains more than 70 different components, including azadirachtin, which accounts for about 90 percent of the oil's insecticidal qualities. It is pressed from the seed of the neem tree (*Azadirachta indica*), native to Southeast Asia. Other compounds include meliantriol, salannin, nimbin, and nimbidin, all of which have fungicidal, as well as other, qualities.

Neem oil is most effective before a major infection begins. It can be used prophylactically for powdery mildew, preventing spores from germinating.

Use neem oil and its derivatives as a spray or as a drench. When used foliarly, it is translinear. That is, it gets absorbed by the leaf and moves around the area. When used as a drench, it kills pathogens and pests in the soil but it is also absorbed by the roots and becomes systemic, moving internally through the plant.

A few brands use cold-pressed or steam-pressed extract of the neem seed. These contain all of the oils found in the seed. Brands include Dyna-Gro, Monterey, Southern Organix, and United Industries.

Clarified hydrophobic extract of neem oil is made using alcohol to remove the azadirachtin, leaving the minor constituents. Some of these phyto-protectors have a more fungicidal effect than azadirachtin, so are the better choice for molds and fungi. Some brands are Trilogy and Green Light Fruit, Nut & Vegetable Spray.

Insecticides that contain only purified azadirachtin and act as pesticides include AzaMax, Bioneem, and Neemix. Insecticides that contain both azadirachtin and extract of neem include brands such as Agroneem.

Neem cake is what remains of the seed after all of the oil has been extracted. It still contains insecticidal, nematicidal, and fungicidal properties, and also has high quantities of N-P-K and micronutrients. In India it is used as a fertilizer. It is applied at the rate of about an ounce per square foot, applied either as a mulch or mixed into the soil.

Peppermint Tea (Indoors) For fungus gnats.

Strong-brew peppermint tea, using one teabag per six-ounce cup of water, kills fungus gnat larvae. Repeat the treatment every three days to kill the larvae as they drop down to the soil soon after they hatch. The tea is harmless to plants and doesn't harm humans or pets.

pH Up (Indoors and Outdoors) For fungi including gray mold and powdery mildew.

pH Up is a generic term for alkaline pH adjustors used to raise water pH in indoor gardens. It is available as either a powder or liquid. Its active ingredient is usually lye (KOH) or potash (K_2CO_3).

Fungi require an acidic environment to grow. They don't germinate or thrive in alkaline environments. Changing the leaf surface environment from acidic to alkaline controls the infection. An alkaline solution with a pH of 8.0 makes the environment inhospitable to the fungus and stops its growth. This is one of the simplest means of controlling fungus. It can be used on critically infected plants as a first step at control.

Potassium bicarbonate (Indoors and Outdoors) For damping off, gray mold, *Fusarium*, and powdery mildew.

Potassium bicarbonate ($KHCO_3$) is very effective against gray mold, *Fusarium*, and powdery mildew because it raises the pH of the leaf surface, making the environment inhospitable to fungal spores. In addition it disrupts the fungal cell's potassium ion balance. Sodium bicarbonate has been used by rose growers for the same purpose for 70 years, but some plants are sensitive to sodium. Potassium bicarbonate leaves a thin layer of potassium, which is a plant macro-nutrient. It is absorbed through the leaf or drops down to the planting medium. Potassium bicarbonate can be used as a preventative, or for severe infections.

See also Baking Soda and Milk, this section.

POTASSIUM BICARBONATE SPRAY

1 teaspoon potassium bicarbonate
wetting agent per directions or ⅛ teaspoon
    castile soap
1 quart water

Mix the potassium bicarbonate and wetting agent in a quart of water. Spray weekly. Potassium bicarbonate is available commercially as Kaligreen, Firststep, and Remedy.

Potassium Silicate. See Silica.

Pyrethrum (Indoors and Outdoors)
For ants, aphids, caterpillars, fleas, fungus gnats, grasshoppers, leafhoppers, mealybugs, root aphids, scales, spider mites, thrips, and whiteflies.

Pyrethrum is a natural broad-spectrum organic insecticide made from the flower heads of the plant *Chrysanthemum cinerariaefolium*, which is closely related to the garden chrysanthemum. It has been in use for over 160 years. Pyrethrum, the extract made from the flowers, contains six compounds called pyrethrins that all have insecticidal qualities.

Pyrethrum is a broad-spectrum insecticide that is toxic to beneficial insects like bees and other cold-blooded animals including fish and reptiles, so it should be used with caution. It is, however, harmless to people and mammalian pets. It kills on contact so it must be sprayed directly on the pests. Pyrethrum is so safe that it is used in flea treatments for dogs, and lice on humans. Confusion exists between the term "pyrethrum" and "permethrin." Permethrin is a man-made insecticide based on natural pyrethrum. It is longer lasting when exposed to light but can be highly toxic to cats.

Pyrethrum is an organic insecticide, but some formulations include a synergistic additive, piperonyl butoxide, a powerful enzyme that interferes with insects' ability to filter out toxins. Including it in the formula helps the pyrethrum stay in the insect's body longer but it is not considered organic.

Pyrethrum is the natural extraction, but there are also synthetic versions called pyrethroids. Pyrethroids use the same pathways, are effective, and low in toxicity. Mammals' metabolic processes quickly dismantle the pyrethroid's toxic effects, although high doses given to rats have produced some damage. Pyrethroids are synthetic compounds and are not considered organic.

Pyrethrum and pyrethroids do not persist in the environment and are quickly broken down by UV light and high temperatures so they do not cause any long-term contamination of soil or surfaces.

Use pyrethrum by spraying it directly on pests. Pyrethrum works by interfering with arthropod nervous systems, disrupting the neurotransmitters. They are paralyzed and die.

Pyrethrum warning: Some people are allergic to natural pyrethrum. One possible indication of sensitivity is if a person is also allergic to similar plants, such as daisies. People with chronic obstructive pulmonary disease, asthma, or any other respiratory disorder should only use pyrethrum when taking precautions to avoid inhalation, or they should consider another product.

A few people have skin reactions such as rashes when they come in contact with chrysanthemums, so gloves or protective clothing are recommended when using this or any type of spray treatment, no matter how "natural." Sneezing is a first noticeable symptom that should not be ignored. If there is any sign of an allergic response to pyrethrum, it is not the right solution for that garden or gardener. Try something else rather than risking adverse health effects.

When it comes to ants, pyrethrum kills quickly, so the ant foragers often don't have a chance to bring it back to the colony. This makes it less appropriate when the goal is to affect the total nest population. Pyrethrum teas poured into colonies or mounds are fairly successful clearing out ants that are immediate problems. Another strategy is to use a lower concentration so it takes the ants longer to die and gives them a chance to carry it back to the nest.

Use the pesticide as a drench to kill insect larvae in the soil. Follow manufacturer's instructions, but use half strength two times a few days apart for fungus gnats and thrips. No treatments should be used for five days before harvesting to allow the pyrethrum to dissipate.

Some pyrethrum products on the market include Bug Buster-O, PyGanic EC 1.4 Pyrethrum, X-Clude Pyrethrum Spray, and Safer Yard and Garden Insect Killer.

Pyrophyllite Clay. See Silica.

Rotenone (Indoors and Outdoors)
Insecticide.
Rotenone is one of the more toxic natural insecticides. It affects mammals and kills not only insects, but fish and amphibians also. It should not be used.

Seaweed (Kelp) (Indoors and Outdoors) For damping off, gray mold, Verticillium wilt, and whiteflies.

Seaweed contains natural hormones and enzymes that are plant biostimulants, as well as a broad array of micronutrients. It is used to stimulate seedlings and reduces damping off. It also makes leaves undesirable to whiteflies. They hesitate to land on areas sprayed with seaweed powder and avoid laying eggs on them.

There are many kelp products available as powders and liquids. Kelp's nutrients and biostimulants are absorbed by the leaves when used foliarly, and they become more resistant to attack. When used as a drench or powder, rates of damping off and rots are reduced dramatically.

Sesame Oil (Indoors and Outdoors)
For aphids, fungus gnats, gray mold, leaf miners, mealybugs, powdery mildew, scales, spider mites, thrips, and whiteflies.
Sesame seed oil has both fungicidal and insecticidal properties. Sesame oil smothers insects

by clogging their spiracles, the breathing holes along their body, and creates an unfavorable environment for fungi to take hold.

RECIPE FOR SESAME OIL SPRAY

1 teaspoon sesame oil. This works out to 1 ¼ ounces per gallon.
1 pint water
⅛ teaspoon lecithin
⅛ teaspoon wetting agent or castile soap

Combine and spray as needed. Sesame oil is often combined with pyrethrum, neem, or soap.

Sesame oil insecticide is available commercially and in mixtures of other oils such as thyme, soybean, and wintergreen oils. Brands include Green Light Lawn and Garden Spray Multi-Insect Killer.

Silica /Silica Salts (Indoors and Outdoors) For gray mold and powdery mildew.

Silica increases plant resistance to stem and bud rot diseases, fungi, and insect attacks; it also increases plant stamina.

Silica is not known to be essential for plant growth. However, when it is available, plants absorb it through their roots. The plants park the silica in the cell wall and inside the cell. They also place it between the cells, to form protective sheaths near the leaf surface. In controlled experiments, plants with high silica content are protected against powdery mildew. Silica is alkaline, so one of its modes of action may be to create a no-grow environment for the fungus.

Plants grown with ample amounts of soluble silica grow thicker cell walls, which result in stronger stems and help resist fungal and insect attacks. Silica also affects the plants' sensitivity to, absorption of, and translocation of several macro- and micronutrients. It also acts as a "toughening agent," increasing the plant's ability to survive stressful situations such as drought, high salinity, and nutrient imbalance.

Hydroponically grown plants cultured with soluble silicon in the growing solutions had reduced incidence and severity of powdery mildew in several trials. It also increased yields and produced thicker, whiter, healthier root systems. When the powdery mildew spore attacks a leaf, silicon is transported to the site where it stimulates the production of antifungal compounds called phenolics that halt the infection process.

There are many additives that can increase silica content to help fight or prevent fungal infections:

- Diatomaceous earth, a fine powder made from the long-dead shells of tiny ocean creatures, is also high in silica. (See Section 1: Diatomaceous Earth.)

- Greensand is a soil conditioner rich in silica. It contains an iron-potassium silicate called glauconite, a trace element amendment and conditioner for soils and potting mixes.

- Perlite also adds silica to soil mixes.

- Potassium silicate is a solution that is used in hydroponics, soil, and foliar applications to provide silica. It contains 3 percent potassium. Brands include Dyna-Gro Pro-TeKt.

- Pyrophyllite clay is an aluminum silicate product in powder form that can be applied as a dust or foliar spray, and is available under a number of brand names (Seaclay, Mineral Magic, Pyroclay).

- Silica stone is a hydroponic medium that can

be used in place of clay pellets and contributes silica to the plant.

- Vermiculite is composed of puffed silica. It contributes silica to the planting mix.

- Zeolite is mined from volcanic areas. It is a combination of aluminum silicates, potassium, calcium, magnesium, iron, traces of manganese, and some tin. Used in everything from aquarium filters to cat litter, this material is marketed as EcoSand, Cline-lite, and ZeoPro.

Sodium Bicarbonate. See Baking Soda.

Sulfur (Indoors and Outdoors)
For damping off, gray mold, insects, powdery mildew, and spider mites.

Sulfur is an effective remedy for plant predators and pathogens including gray mold, powdery mildew, insects, and mites. It is probably the oldest known pesticide/fungicide still in use,

referred to by Homer nearly 3,000 years ago as "pest-averting sulfur." Sulfur prevents fungal spores from germinating by lowering the pH of the leaf surfaces, so it is an excellent preventative as well as a solution. It is a contact pesticide so sprays and sulfur vapors are excellent means of application.

Sprays are most effective before the plants are affected, or under light infection. They should not be used with oils or when the temperature is over 85 °F (30 °C). Foliar sulfur sprays sometimes cause leaf damage.

The preferred method of delivery is vaporization. Heat elemental sulfur or garden sulfur in a sulfur burner, available in indoor garden shops and on the Internet. Sulfur is heated using a 60-watt incandescent light bulb that vaporizes the element. The vapor condenses into a fine film of sulfur particles on leaf surfaces. These granules have a low pH that inhibits fungal growth. Sulfur candles work in the same way.

Sulfur sprays are also available, but may cause leaf damage, so try any potential controls on a few leaves before spraying the whole plant. Sulfur produces a strong smell as it vaporizes, so air out the growing space before reentering. Do not use oils and do not use when the temperature is over 85 °F.

Elemental sulfur, also called garden sulfur, is available as Thiolux Jet and Yellowstone Brand Hi-Purity Prill.

Sulfur burners that emit vapors are often used to control powdery mildew.

UVC Light (Indoors and Outdoors)
For bacteria, damping off, fungi, all molds, powdery mildew, and root rot.

UVC light is considered deadly to life and kills the spores and tissues of powdery mildew, *Pythium spp.*, *Botrytis*, and other fungal pathogens.

There are three ways that it is used.

First, there are water sterilizers that kill pathogens in the water supply. As the water passes around the UVC fixture, microorganisms receive fatal rays. The fixtures can also be used to clean inline water in recirculating hydroponic systems after filtration.

Place the fixtures in the ventilation systems of growing rooms to kill fungi and bacteria spores before they enter the growing space. In closed areas use systems designed for sterilization in restaurants. The light is fatal to all airborne organisms passing through it. Some brands of lights are Big Blue, Turbo Twist, and Air Probe Sanitizer.

Hand-held UVC wands and automated systems are used to prevent fungal infections. They kill powdery mildew spores on leaf surfaces. The light passes over each plant for only a second each day. This is enough to keep the plants fungus-free. Aeon UVC Systems is the only company manufacturing for the garden industry.

Vinegar (Indoors and Outdoors)
For powdery mildew.

Vinegar is an acid and is toxic to powdery mildew. Use 1 tablespoon per quart of water and spray on foliage as a control. Alternate with potassium bicarbonate and milk for more complete control. Apple cider vinegar is a good choice.

Wood Ashes (Outdoors)
For damping off, caterpillars, and snails and slugs.

Wood ash is useful as a fungicide and pesticide because it is strongly alkaline. This inhibits growth of fungi. Insects which cross over it get burned from its high reactivity. Sprinkle it sparingly around the plants but not closer than 2 inches from the plant's stem.

SECTION 3:
Biological Controls

Biological controls consist of either living organisms or products they produce. They range from microbials such as fungi and bacteria, to tiny parasites and predators. Some biological controls are carnivorous insects and arthropods with a hearty appetite. These parasites often mature inside the victim. They all have different means of earning a living—killing pests—but are very effective. Biological controls have advantages over any kind of chemical or herbal pesticide because they are harmless to humans and pets and most can be used anytime during the plant cycle. Many reproduce, offering continuing protection.

Biological controls are purchased as both live organisms as well as biological-based products. Some sources are Arbico Organics, Tip Top Bio-Control, Koppert, Rincon-Vitova, Nature's Control, and GreenMethods.com.

MICROBIAL PESTICIDES (MPs)

Beneficial bacteria, nematodes, viruses, and fungi are microbial pesticides. MPs are for pest and disease control, using "inundative release." Millions of individual nematodes, spores, or bacteria are released to attack an infestation, just as millions of droplets of chemical pesticides are used to quell an outbreak in conventional pest control.

Most MPs must come in direct contact with the pest to work. Bacterial MPs infest the host—the target pest—then produce antibiotics to ward off competition from other bacteria. Once they have the host to themselves, they produce enzymes that break down the cell walls of their hosts.

Fungi attack plants by growing hyphae that invade the target's cells, sometimes producing poisons or enzymes to disarm the victim and digest it. Nematodes invade plants through wounds or holes in the structure and then release bacteria that digest the tissue. Then the nematodes eat the vegetative-bacterial broth.

MPs formulated for soil application work best if applied during a heavy rain outdoors, or watered well indoors. Most MPs should be applied with non-chlorinated water, spray oils, wetting agents, or other spray adjuvants such as UV protectors or feeding attractants.

Beneficial nematodes are effective against some insects. Some biocontrol agents, such as ANTidote for ants, are composed of more than one kind of nematode. Others, such as *Phasmarhabditis hermaphrodita*, are effective against snails and slugs, and *Steinernema feltiae* attacks several insect species.

PARASITES AND PREDATORS

Parasites are used to control or eliminate pests or diseases before the outbreak has reached epic proportions. Predators are recommended for heavy infestations. However, this may just be a prejudice caused by the subtlety of parasites as compared to the aggressive moves of the predators.

Predators seek out insect victims, then kill, and eat them. Parasites can inject an egg into the insect. It hatches and the larva matures inside the insect and eats its way out, or uses the target as a continuing source of nourishment. Some species of nematodes and fungi use this technique.

Parasites

Parasites kill their prey from within. Their usual mode of operation is to insert one egg each into multiple hosts. The egg hatches inside the host and it consumes the host alive, usually leaving vital organs for last. The parasite then pupates inside the host's cadaver and emerges as an adult. Some species parasitize in two ways: The larvae feed on the hosts they were born within, and the emerging adults consume many more hosts.

Parasitoids tend to stick around an area they have been introduced and will voraciously hunt their prey until nearly all are gone. They are crop or pest specific. As the adults emerge from the pupae they are imprinted with the odor of the specific crop, causing adults to remain on that crop and search for that pest instead of searching for prey on other crops.

Examples of parasites are the wasps *Encarsia formosa* (whiteflies), *Trichogramma* wasps (caterpillars), and *Aphidius colemani* (aphids).

Predators

While not as dedicated as parasites, predators use a more direct method of controlling pests.

Unlike parasites, predators consume more than one pest before reaching adulthood. They do not incorporate their prey into their life cycle, but instead use them for food. Most predators have either chewing mouthparts or piercing-sucking mouthparts.

Some predators are general consumers. They catch and eat a large variety of pests. Unfortunately, this may also include other beneficial insects. Other predators, such as lacewings and ladybugs, are more selective in their diet.

Most predators eat the largest insects they can attack first, no matter that they may be surrounded by other, smaller prey. For instance, predators introduced to eat mites who also have an appetite for caterpillars go for the caterpillars first every time. This is why selective predators such as predatory mites work best for a particular pest. They are genetically programmed to dedicate their efforts to specific pests.

Some predators go into hibernation when the days grow short. Others fly away when the number of available prey falls. Gardeners can entice predators to stay with alternative food sources, such as pollen, artificial honeydew, or water.

BACTERIA
Bacillus pumilus (Indoors and Outdoors)
For powdery mildew and rust.

Bacillus pumilus QST 2808 strain produces an antifungal amino-sugar compound. This compound disrupts cell metabolism and destroys cell walls, leading to the destruction of the cell and death of the plant pathogen. *B. pumilus* also inhibits pathogen growth at the leaf surface so it cannot obtain a foothold into the plant tissue.

B. pumilus is effective for about 10 days. I found that it does not completely eliminate powdery mildew when used alone, but it does work very well in conjunction with *B. subtilis*.

B. pumilus is available as Sonata.

Bacillus subtilis (Indoors and Outdoors)
For damping off, *Fusarium*, gray mold, powdery mildew, *Pythium*, *Rhizoctonia*, root rot, *Sclerotinia sclerotiorum*, and Verticillium wilt.

Bacillus subtilis colonizes the soil around plant roots and produces antibiotics that suppress the fungi causing root and stem rot. *B. subtilis* is used as a seed treatment or soil drench. It works better as a preventative, and will not control a serious damping off infestation after it has begun.

The *Bacillus subtilis* strain (QST713) is known by the brand name Serenade. This strain is particularly effective for powdery mildew. It uses three chemical pathways to destroy disease-causing pathogens. It stops harmful spores from germinating, disrupts growth of the germ tube and mycelia, and inhibits the growth of the fungus at the leaf surface. It is considered totally safe to humans and animals since the bacteria attacks only fungi. Watch out if you are a fungus. Otherwise you are safe.

This bacterium is marketed under several names, including Companion and Serenade.

Bacillus thuringiensis (Indoors and Outdoors) For caterpillars, fungus gnats, grubs, and larvae.
The bacterium *Bacillus thuringiensis*, or Bt, was first described by Louis Pasteur in the 1860s, but was not tested in the field as a natural pesticide until the 1920s. In the 1950s, it became available for use extensively for the control of corn and tobacco caterpillars. Bt is a family of bacteria

with at least 35 separate strains producing 140 types of spore toxins, the real weapons against pests. Two common Bt strains are described here because they contain spore toxins that are active against some common cannabis pests, as well as beetle grubs, fly larvae, and many other soil-dwelling insects.

Bt-i (variety *israelensis*) (Indoors and Outdoors) For fungus gnats and mosquitoes.
This spore toxin exists in the bacteria as a crystalline protein. When a fungus gnat ingests a spore, the protein is dissolved by enzymes in the insect's digestive tract and destroys the cells of the insect's digestive system. Once the spores are ingested, insects stop feeding within an hour. They shrivel, blacken, and die. Bt-i does not reproduce in pest populations, so it must be reapplied every week to 10 days. The dust is applied to the surface of the soil or rockwool. Bt-i is compatible with some other biocontrol agents, including the beneficial nematode *Steinernema sp.* and the soil mite *Hypoaspis miles*.

It is marketed as Gnatrol and Bactimos.

Btk (variety *kurstaki*) (Indoors and Outdoors) For caterpillars.
Btk (*Bacillus thuringiensis*, var. *kurstaki*) is one of the best solutions for caterpillars. Btk is a naturally occurring soil bacterium that has been used since the 1930s. It acts selectively, killing caterpillars, but poses no harm to beneficial insects, earthworms, fish, birds, cats, dogs, other mammals, or humans.

Btk renders the caterpillars' stomachs nonfunctional by multiplying inside the digestive tract, creating sharp, toxic, protein crystals. Btk does not attack *Lepidoptera* eggs, but it works well on caterpillars that are chomping on leaves, especially if they are in the earliest stages of caterpillar life. Once the leaves have been sprayed with Btk, only caterpillars that eat the treated areas will be affected. Btk affects insects quickly, causing caterpillars to stop eating soon after they've fed on treated plants. Death follows in a day or two. The decaying caterpillar body releases more of the bacteria.

Given the lengthy commercial use of this Bt strain, there is some concern about caterpillar resistance. There are no known incidences of developed resistance to Btk among caterpillars.

Btk comes in liquid and wettable powder form. A single application provides permanent protection until it is washed off, so reapply after rain. Btk breaks down over three to seven days. Since this product works best on young caterpillars, multiple treatments are necessary to eradicate all caterpillars from staggered hatch times.

If product instructions call for dilution with water, use nonchlorinated water because chlorine in tap water can destroy the Btk bacteria, thus rendering the treatment useless.

Btk is widely available in garden stores and through the Internet. Brands include Green Step Caterpillar Control, Dipel Dust, and Bonide Bt Thuricide. Make sure that the ingredient label of any other brand does not list additional chemical pesticides.

Saccharopolyspora spinosa (Indoor and Outdoors) For ants, caterpillars, leaf miners, spider mites, thrips, and other arthropods.

This bacterium produces an insect toxin, spinosyn, when it is cultured in a nutrient broth. This rare but naturally occurring bacterium was sup- posedly discovered in an abandoned rum distillery in the Caribbean in 1982 by a scientist who collected soil samples hoping to discover a new organism with natural insecticidal properties. These samples contained the bacterium *Saccharopolyspora spinosa*, which produces metabolites that are injurious to insects. Since that time, *S. spinosa* has not been found again in nature.

A new class of organic insecticides is produced from two of the fermentation products of this bacterium: spinosyn A and spinosyn D. In combination, these metabolites act by highly exciting the insects' nervous system to the point of dysfunction. Mortality is assessed at 100 percent. This combination acts when ingested and on contact. It does not affect non-target insects, such as beneficials. It is toxic to bees, however, so use outdoors with caution. It acts quickly; insects die within one to two days, but it must be eaten by them to be effective.

It is marketed under the active ingredient name spinosad. Spinosad (spinosyn A plus spinosyn D) does not persist in the environment and breaks down into carbon, hydrogen, oxygen, and nitrogen when exposed to sunlight and microbes.

Brands include Monterey Garden Insect Spray, Captain Jack's Deadbug Brew, Conserve SC, and Entrust.

Actinomycetes are a group of filamentous (rod-shaped) anaerobic bacteria that have qualities that somewhat resemble fungi. Their phenotypes are dependant upon the environment. The three mentioned here (*Saccharopolyspora spinosa*, *Streptomyces griseoviridis*, and *Streptomyces lydicus*) are mobile, using their flagella—whip-like tails—to propel themselves.

Streptomyces griseoviridis (Indoors and Outdoors) For damping off, gray mold, root rot, and Verticillium wilt.

This strain is an actinomycete, a member of a curious group of organisms that fall in between fungi and bacteria based on gross morphology, but are truly bacteria. This particular organism is native to the peat bogs of Finland and colonizes the surfaces of plant roots. As it grows, it produces metabolites that inhibit the growth of *Fusarium*, *Botrytis*, *Rhizoctonia*, and *Pythium*, and enhances the host plant's growth. Wonderfully accommodating, *S. griseoviridis* prefers humid soil, a wide range of soil pH levels, and cool soil temperatures (between 50 and 68 °F). It is also compatible with other growing additives, such as rooting hormones, and does not seem to affect beneficial organisms in the soil.

S. griseoviridis is available as spores and mycelial fragments in a powder. It grows a protective sheath around roots protecting them from attacks by pathogens. It also forms a bond with the roots that results in increased vigor and stress resistance. Use it to treat seeds, as a root dip, or soil drench. It is marketed as Mycostop.

Streptomyces lydicus (Indoors and Outdoors) For damping off, gray mold, powdery mildew, root rot, Verticillium wilt, and other pathogenic root fungi.

Streptomyces lydicus is a naturally occurring common soil bacterium normally present in healthy "living" soil. When it is inoculated to improve poor soils and planting mixes, it provides protection against pathogenic fungi. It grows on the tips of plant roots, attacking pathogenic fungi. It can also be used as a foliar spray to treat powdery mildew and gray mold. *S. lydicus*

grows well in normal garden temperatures and pH levels.

The bacterium spreads along the plant's roots as they grow into the rhizosphere, which is the area around the root. This shields plant roots from pathogens. The bacterium releases an enzyme which weakens the fungus' cell wall as well as releases fungicides that inhibit the pathogen's growth.

Use in early growth so it establishes early. *S. lydicus* does hinder mycorrhizal fungi growth.

Hydrate the powdered bacteria by mixing it into soil mix. To use as a drench or foliar spray, mix it in water. It is safe for use around insects, other animals, and people.

In hydroponic or sterile systems, introducing *S. lydicus* restores its natural presence on plants. In soil systems, adding *Streptomyces lydicus* elevates colony levels for increased benefit.

It is available under the brand name Actinovate SP. *Streptomyces griseoviridis* (available as Mycostop) is a similar bacterial strain used for similar purposes.

BEETLES
Cryptolaemus montrouzieri (Indoors and Outdoors) For mealybugs.

Mealybug destroyer feasting.

Popularly known as a mealybug-destroyer, *Cryptolaemus montrouzieri* is a lady beetle that preys on many species of mealybugs, including both the citrus and long-tailed. It will also eat some scales, although they are not its preferred diet.

Cryptolaemus females deposit eggs in mealybug population centers near the mealybug egg sacks. The *Cryptolaemus* larvae, which resemble mealybugs in that they are white and carry waxy appendages, emerge in about five days and immediately start chomping on mealybug eggs and crawlers. *C. montrouzieri* goes through three larval stages, molting each time. The time required in each of these stages is determined based on temperature. Crypts (the larvae) can last anywhere from 20 days in cool weather, or speed up to a life cycle of 12 days when it is warm. As the young beetles grow, their taste expands to include mealybug adults.

The crypts search for sheltered places on the leaves and branches, weave a cocoon, and pupate. In one to two weeks the winged adults break out of the cocoon after a complete metamorphosis. They no longer look like a mealybug. The adults are about 4 millimeters long, brown/black with an orange head and thorax. They live for a month or more, continuing to munch mealybugs. A females lays about 500 eggs in her lifetime.

Lady Beetles (Ladybugs) (Outdoors)
For aphids, mealybugs, scales, spider mites, thrips, and whiteflies.

Adult lady beetles (*Delphastus pusillus*) are 4 to 7 millimeters long. Their bright coloring serves as a protective shield, indicating danger, and alerting predators that they are toxic. When threatened they emit an alkaloid poison from their joints.

If the temperature stays moderate and there is an ample supply of aphids, they will continue to hang out and lay eggs until late summer. Then they migrate to the highlands for the winter.

Lady beetles converging on prey.

Adult lady beetles overwinter in big aggregations, selecting sheltered areas under leaves or rocks, or inside structures. They often migrate hundreds of miles to hang out together. When it warms up in spring, they disperse and search for food. Then they mate and lay eggs. The adults continue feasting and laying eggs over a three-month period.

Lady beetle eggs are cream or yellow in color and about 1 millimeter long. The eggs are laid in clusters on leaves, near prey. The number of eggs laid and their fertility depends on food abundance, ranging from 20 to 1000. When they hatch, the newborns are less than 1 millimeter in length. They immediately search for food. The moms lay infertile eggs for the hatchlings to eat as their first food.

The larvae look a lot like crocodiles. They molt three times into four instars, growing to 1 centimeter in a month. Then they spin a cocoon attached to a leaf and pupate for three days to two weeks depending on species and temperature. The pupae are often brightly colored yellow and orange. Newly emerged adults look for food and sex.

The Many Lady Beetle Varieties

There are over 450 species of lady beetles native to North America. These include natives such as the convergent lady beetle (*Hippodamia convergens*), the most common ladybug, which is ubiquitous throughout the continent. *H. convergens* is named for two yellow stripes that converge on its thorax, right behind its head.

Adult lady beetles are carnivorous and eat up to 10 to 50 aphids a day. Some ladybugs have a specialized diet. This is often based on nutritional requirements due to their size. Others, such as the *H. convergens*, are general feeders with preferences for aphids. However, they will eat any prey they encounter, including thrips and scales.

Harmonia axyridis, the harlequin ladybug, is native to Asia. Several unsuccessful attempts were made to adapt it to North America for its voracious appetite for whiteflies. In 1988 a population was discovered near New Orleans that resulted from an unplanned import. Since then it has spread throughout North America and Europe. This is the species notorious for invading selected homes in the fall.

All but two species of ladybugs are carnivores. These exceptions are the Mexican bean beetle (*Epilachna varivestis*), and the squash beetle (*Epilachna borealis*), which look a lot like their carnivorous cousins but are voracious herbivores. They are not a threat to cannabis. The species can be identified by the eight black spots of variable size on each wing cover, arranged in three longitudinal rows, and their bronzed orange-brown color.

The lady beetle, *Ryzobius lophanthae*, is a voracious soft scale predator. A clutch of eggs is laid underneath a scale infestation. The dull gray larvae feast on scales for three weeks then metamorphosize into tiny black fuzzy ladybugs with orange heads and thoraxes. The adults live for six to eight weeks eating scales. The females lay hundreds of eggs when the scale population is large. *R. lophanthae* is opportunistic and also eats the aphids and mealybugs that it encounters. However, it is most effective on scales.

Convergent lady beetle (*H. convergens*).

Multicolored Asian lady beetle (*H. axyridis*).

Multicolored Asian lady beetle (*H. axyridis*).

Ladybugs eating aphids.

Once released, their dispersal instinct takes over and they fly away. Twelve hours after release, only the weak, infirm, or confused remain in the garden.

Attraction is a far better strategy. Lure them to your garden. Water droplets for drinking and a source of pollen—especially mustards—to round out their diet are good lures.

Lady beetles are top feeders. They look for plentiful prey. When the food source becomes scarce, they fly to new sites with more food.

H. convergens is the species available commercially. Adults should be released at sundown. In larger gardens, release them weekly. The beetles will dine on any prey before they move on.

Another strategy is to introduce the beetles only when an infestation is evident. Adult females must consume the equivalent of more than 100 whitefly eggs per day to lay eggs. When the whitefly population is thin, ladybugs seek other prey including spider mites and aphids. Although they are top feeders, usually leaving the small portions found lower on the plants for others, they will search out food if they cannot leave, such as in a greenhouse. Wheast can be used as a substitute for pollen, which they require as part of their diet.

Rove Beetles (*Atheta coriaria*) (Indoors and Outdoors) For fungus gnats and thrips.

A. coriaria, the rove beetle, is used for the biological control of fungus gnats. It is black, 3 to 4 millimeters long, and slender with short wing covers. It comes in various brown shades. These beetles are soil dwelling. Eggs hatch in four days, followed by three larval stages that look like small versions of the adults, varying in color from white to yellow-brown. The larvae feed for about two weeks and the pupal stage lasts three to four days.

Rove beetle.

Both adult and larval stages are predatory. This beetle establishes a permanent population easily, with abundant prey. For a more effective treatment combine *A. coriaria* along with the predatory mite *H. miles*.

A. coriaria comes as adults in shaker bottles. Apply about 150 per 10 square feet of space. Water the soil no more than 10 minutes before release and then not again for about two hours. Rotate the container to loosen the beetles from the sides, and then shake them out onto growing media, algae-thick areas, and under greenhouse benches. Leave the empty container in the garden for 24 hours to assure all the beetles have exited the container. Do not store the beetles for more than four days and keep the bottles in humid conditions at temperatures of 45 to 55 °F prior to release.

BUGS
Big-eyed Bugs (*Geocoris pallens* and *Geocoris punctipes*) (Outdoors) For aphids, spider mites, and whiteflies.

The two most common species of big-eyed bugs

are found in gardens all over North America. At about 3 millimeters long, these little critters enjoy munching on aphids, mites, and other harmful invaders in the garden. They have oval bodies and broad heads with distinctive, wide-set bulging eyes and short antennae with enlarged tips. They have two sets of wings. Adults are gray, brown, or black in color. They look a bit like common flies. Nymphs look similar to adults but don't have wings.

Big-eyed bug.

Big-eyed bugs are true bugs; that is, they have a piercing-sucking mouthpart. Once a bug has grabbed an aphid or other insect securely, it pierces its prey with its straw-like mouth and injects digestive juices that turn the flesh to liquid. Then it drinks its high-protein smoothie. This method of ingestion allows the bug to eat prey larger than itself. Big-eyed bugs are generalist feeders with big appetites, eating aphids, whitefly eggs, mites, and any other slow-moving prey they can catch.

The female lays pinkish, oblong eggs near prey that hatch in five to 10 days. The nymphs are bluish-gray with prominent eyes, and look like adults without wings. They go through five nymphal molts, and eat prodigiously to support their rapid growth. Like the adults, the nymphs' diet is composed of insects.

Both nymphs and adults eat honeydew and leaf parts if other food is scarce. The wingless nymphs don't travel far. After eating up to 300 mites or dozens of aphids, nymphs mature into sexually active adults in about 30 days. Adults live up to three months, continuing to sip liquefied pest flesh. The female lays about two eggs a day throughout her life. Although they play a major role in plant production, only a few insectaries offer big-eyed bugs.

Damsel bugs (*Nabis alternatus*)
(Outdoors) For all insects they can grasp, including caterpillars of all kinds.

Damsel bugs, or nabids, are members of the order Hemiptera, true bugs. They are general predators of soft-bodied insects. Adults are 8 to 12 millimeters long. They are tan to reddish-brown; have flat, elongated, slender bodies; stilt-like legs; curved, needle-like beaks; and look like a cross between a praying mantis and a stink-bug. Their raptorial front legs are used to grasp prey. They hold these legs up as if they are delicately lifting a skirt, earning them their common name, "damsel bug." Both larvae and adults eat a large variety of insect pests, including whitefly larvae and pupae.

Damsel bug nymph.   Damsel bug adult.

Like all Hemiptera, their life cycle is a simple metamorphosis. Females insert their eggs into plant tissue with their ovipositor. Nymphs hatch, pass through five instars, and molt into adults in about 50 days. Damsel bugs are found in the wild and overwinter well, even in colder areas, so they are suited for use in outdoor growing areas. They are also available from insectaries.

Minute Pirate Bugs (*Orius sp.*)

(Indoors and Outdoors) For aphids, potato-type leafhoppers, spider mites, thrips, and whiteflies.

Minute pirate bug puncturing an aphid.

The minute pirate bug is small, only ⅛ to $1/10$ of an inch (1.5 millimeters) long, so it picks appropriately sized victims. Thrips and spider mites are its preferred food, but it also enjoys sucking the sweet flesh of small aphids, insect eggs, an occasional caterpillar, or anything else it can catch. Once the thrips population is under control, pirate bugs survive on aphids, mites, scales, whiteflies, other biological controls, and each other.

The minute pirate bug (*Orius sp.*) is a true bug: Its mouth is modified into a sharp hollow beak through which it slurps its dinner. The bug catches its prey, holds it with its two front legs, and uses its beak to pierce the exoskeleton. Then it sucks the victim's flesh through its built-in straw. Unlike some bugs that inject a protein liquefier, the pirate bug seems to use mostly suction power to drink its dinner. Usually it pierces the prey in several areas to get at different sections of the insect. When it has finished, only the prey's exoskeleton is left.

The minute pirate bug life cycle starts with an oval egg approximately $1/100$ of an inch long, deposited in a small hole in a leaf that the female pokes. The egg hatches in three to five days and the nymph, also only $1/100$ of an inch long when it hatches, starts hunting for food. Nymphs are bright yellow to brown with red eyes and are teardrop-shaped. Over a period of 20 to 25 days, the nymph undergoes five instars, similar to molts. It emerges as an adult with two sets of usable wings. Now that it has transportation, the bug starts searching for mates. Three days after the female has mated, she starts laying eggs. Over her 35-day lifespan, she lays an average of about 130 eggs, almost four per day. Development to reproductive maturity takes about two weeks and adults live for about a month. Adults are long and vary from yellow-brown to black with white wing patches. They overwinter as adults and produce two to four generations.

Combining Forces: Thrips went unnoticed until their telltale leaf etchings were discovered. By that time it was a deep infection. A single release of *O. insidiosu*s at the rate of one per square foot, along with the release of beneficial nematodes, eliminated the pests in about 10 days. The pirate bugs feasted on the adults up in the canopy, while the nematodes enjoyed the succulent pupae that dropped from the plant to the soil to phase-change into adults.

Minute pirate bugs have several idiosyncrasies that are important to know about. First, they are cannibalistic. If the bugs are kept together for long, eventually a few fat individuals will emerge.

Secondly, they undergo diapause (similar to hibernation) under a long night cycle. This means they are effective in vegetative rooms but not in flowering rooms because they become inactive under the long night regimen. Cannabis plants regulate their photoperiod response based primarily on red light, but also slightly on blue. Minute pirate bugs are sensitive to blue light in determining diapause. Use a pure blue light using LEDs, or blue fluorescent tubes, to stop them from hibernating. Shield the light from the plants, since it will interfere with the flowering cycle.

Third, minute pirate bugs eat pollen as well as flesh and prefer to be somewhere where they have access to it. Outdoors, it is a good idea to grow flowers that provide a ready source of pollen. Indoors, if there is no pollen for them to eat, Wheast keeps the bugs content and encourages egg laying. The bugs also partake in an occasional sip of plant juice. This causes the plant no harm.

Fourth, minute pirate bugs bite. You wouldn't think that such a small insect with no venom could inflict the sharp pain they do, but when they pierce the skin it hurts. Luckily this does not happen often, and they don't inject a toxin or anti-coagulant so there is no lingering pain or swelling.

Six different species of minute pirate bugs are available for thrips control. They prefer temperatures of 68 to 86 °F and relative humidity of 60 to 85 percent. Two species are *Orius tristicolor* and *Orius insidiosus*. *O. tristicolor* is usually found in western states while *O. insidiosus* is more common in the Midwest. It works well for potato leafhoppers. *Orius* species are important but usually unseen predators in fields and gardens. They search for thrips and mites and protect pastureland, orchards, and field crops. Vegetables and berries often have large herbivore populations and *Orius* species help reduce them.

O. insidiosus is available commercially for greenhouses and grow rooms. Since they are cannibalistic it is important to get the bugs into the garden as soon as they are received. They should be released in the evening in small groups around the infected growing areas.

They are shipped as adults and once released will fly short distances around the indoor garden seeking prey, mating, and laying eggs. Since they lay a lot of eggs and can complete a cycle from egg to egg-laying adult in less than a month, *Orius* populations build up quickly, outpacing the herbivores' reproductive abilities. Adults introduced to the garden will immediately lay eggs, which will hatch within a week of dispersal.

Outdoors, there is probably an *Orius* species already in your garden or near it. You can encourage them to set up shop on your property by growing flowers that produce a lot of pollen, which they eat as a side dish with their flesh shake. Most annuals and many perennial plants produce clusters of pollen-producing flowers.

Orius species can do a lot of damage for their small size. They eat 30 mites or thrips a day, even as young nymphs, and they are sloppy at it. A minute pirate bug might be in the middle of a mite meal, get bored, leave its victim half eaten, and instead capture another living morsel and begin a new entrée. Over its lifetime of less than two months, a single minute pirate bug will consume about 1,500 plant pests.

Orius species and prey

Species	Preferred Prey	Co-exists With	Diapause
Orius tristicolor	Thrips (*F. occidentalis, T. tabaci, Caliothrips fasciatus*), spider mites, aphids, budworm eggs	Mite *N. cucumeris*	Daylengths < 13 hours
Orius insidiosus	Thrips (*F. occidentalis*), spider mites, aphids, budworm eggs, aphid predator *Aphidoletes aphidomyza*	*N. cucumeris*	Photoperiods < 11 hours Warm temperatures may postpone
Orius albidipennis	Thrips (*F. occidentalis, T. tabaci*), spider mites, aphids, whiteflies, budworms, cutworms	Lady beetles, lacewings, *Orius laevigatus*	Does not diapause
Orius laevigatus	Thrips (*F. occidentalis, T. tabaci*), spider mites, European corn borers, aphids, cutworms	Lady beetles lacewings, *Orius albidipennis*	Photoperiods < 11 hours Warm temperatures may postpone
Orius majuscules	Thrips (*F. occidentalis*)		Photoperiods < 16 hours Warm temperatures may postpone
Orius minutes	Thrips (*T. tabaci*), spider mites, European corn borer eggs, aphids		Daylengths < 13 hours Warm temperatures may postpone

Outdoors, *Orius* species are active from spring until fall when they go into diapause, and wait out the winter in fallen leaves and other plant litter. They become active again in the spring, shortly after the plant pests emerge. *Orius* species are sensitive to chemical insecticides, which severely reduce their numbers, but they are not sensitive to the biological controls used for thrips, mites, or aphids.

Commercially supplied as adults and nymphs, they come packaged in shaker bottles and can be stored for up to three days in a cool 46 to 60 °F, dark place. Release one predator per plant or one predator per square foot for moderate infesta-tions. Release five to 10 predators per square foot for heavy infestations. Outdoors, adults fly away when food becomes scarce. They prefer flowering plants and eat the largest pests first. All species are sensitive to pesticides.

Anthocorius nemorum is a minute pirate bug closely related to *Orius* and is used against thrips in England.

Mirid Bugs (Indoors and Outdoors)
For thrips and whiteflies.

Sold only in Europe and not commonly found in gardens, Mirid bugs are the largest group of true bugs in the suborder Heteroptera. They have iconic craggy angular legs and narrow defined bodies. There are over 10,000 species that vary in shape and color, which is mottled and ranges from drab to bright colored. They reside in grass, on leaves of trees, and on plants. Several types of mirid bugs are beneficials and are predators of pests, especially thrips and whiteflies.

Mirid bug (*Deraeocoris brevis*).

Deraeocoris brevis mirid bug eats all stages of thrips. They live in temperate climates and are active when the temperature is 64 to 86 °F and the humidity is 30 to 70 percent. Adults are 5 millimeters long and brown with mottles on their backs. The nymphs are a lighter color of brown and look like miniatures.

Marcolophus caligenosus is a form of mirid bug native to the Mediterranean. The release of *M. caligenosus* is currently prohibited in North America. They are sold in Europe to control whiteflies and also consume spider mites, caterpillars, and leaf miners.

Chalcid wasps. See Wasps, Parasitic: *Trichogramma* wasps.

Cryptolaemus montouzieri. See Beetles.

Delphastus pusillus. See Beetles: Lady Beetles.

Deraeocoris brevis. See Mirid Bugs: *Deraeocoris brevis*.

Compost Tea. See Part 2: Compost Tea and Chapter 1: Compost Tea.

FLIES (SEE ALSO: MIDGES)
Coenosia attenuate (Indoors and Outdoors) For fungus gnats.

C. attenuate, known as the "hunter fly" or "tiger fly," is related to the housefly. This unremarkable-looking grayish fly is an avid predator of fungus gnat adults and other flying insects. Its larvae live underground and are generalist predators on soil insect larvae, including fungus gnats.

Phorid fly (Pseudacteon) (Outdoors) For ants.

Phorid flies are currently in use by federal and state authorities for the control of fire ants in public lands, but are not yet available for home garden use.

FUNGI
Beauveria bassiana (Indoors and Outdoors) For ants, aphids, some beetles, caterpillars, fungus gnats, grasshoppers, root aphids, spider mites, termites, thrips, whiteflies, and other arthropods.

This fungus parasitizes a wide variety of insects and can be used on vegetables and other food crops. It is safe for humans and mammals but it is toxic to beneficial insects.

B. bassiana is a naturally occurring parasitic fungus found worldwide. *Beauveria* species are not as lethal as *Verticillium lecanii* (see below in this section), but have a broader range of hosts. *B. bassiana* causes white muscardine disease. Three to seven days after *B. bassiana* spores come in contact and stick to an insect, they germinate, and penetrate its body. Then they grow inside. Infected insects stop eating soon after and die in two to 10 days. The fungus' reproductive organs emerge from the corpse as a white mat (hence the name white muscardine disease) and produce spores. In caged rose trials, *B. bassiana* killed 82 percent of thrips.

B. bassiana does best at 90 percent humidity but works well in the grow room humidity range of 40 to 60 percent, with daytime temperatures of 70 to 86 °F. Spores do not have to be ingested to work.

While harmless to humans and pets, *B. bassiana* is harmful to many beneficial insects such as green lacewings, lady beetles, and other soft-bodied insect predators. Some people have developed allergic reactions to *B. bassiana* upon repeated exposure. The fungus establishes in the soil, but mixing it with a large proportion of organic matter hinders its growth and high rates of nitrogen fertilizer kill it.

Spray *B. bassiana* on all surfaces of infested plants and the medium. Once it establishes in a space, it helps to keep infestations from occurring. *B. bassiana* comes as spores in vegetable oil, or water-dispersible granules, and may be stored for up to two years in a cool (46 to 50 °F) and dark place. Some strains have some residual activity but, generally, the mixture acts as a contact insecticide. Adding neem oil increases the effectiveness.

Brands including BotaniGard ES and My-cotrol O contain the GHA strain of *B. bassiana* specified for aphids, mealybugs, thrips, and whiteflies.

Mycorrhiza (Indoors and Outdoors)
For damping off, gray mold, root rot, general plant health, and nutrient deficiencies.

The term mycorrhiza is used to describe the root in combination with the fungus. The literal translation is fungus (*myco*) root (*rhizo*). The fungi that form mycorrhizae are called mycorrhizal fungi.

Although fungi like gray mold, damping off, and root rot are all destructive pathogens that are a cause for alarm, not all root fungi are harmful to plants. Mychorrhizal fungi are beneficial and present in most soils naturally, especially soils high in compost. Inoculants such as Bountea Root Web, MycoGrow, Great White, or Rick's Monster Grow can be added to ensure a healthy colonization. At least 90 percent of plants regularly form this relationship with mycorrhizal

Bountea's Root Web is an endo/ecto mycorrhizae, with 19 species of beneficial fungal spores aiding plants' root growth, bolstering fertilizer efficiency, and improving plant yields.

fungi. Orchids for example cannot survive the seedling stage without the additional help that mychorrhizal fungi provides.

Mycorrhiza fungi are symbiotic with the plants they connect with. Functioning somewhat like "root extensions," the fungal strands are smaller in diameter than the plant roots, and allow the plant to benefit from a wider surface area. As the fungus spreads through the media, the network of strands can increase resources available to the plant roots hundreds or even thousands of times. Plants with mycorrhizal roots are better fed, more drought resistant, and have a higher resistance to pathogenic infections. Another benefit provided is the filling of their environmental niche in the ecosystem, taking resources that otherwise may be available to pathogenic fungi.

The fungi attach to the plant's roots, feeding on the plant's carbohydrates. In return, the fungi supply the plant with moisture, nitrogen, phosphorus, copper, calcium, magnesium, zinc, and iron, which the fungus collects and makes available to the plant.

Since the fungus consumes carbohydrates for food, carbohydrates (sugar, molasses, etc.) are sometimes added to nutrient solutions to improve the colony growth. Plants don't absorb carbohydrates directly, but the resulting additional vigor in the mycorrhizal and microbial life can improve the health of the plant, and the size and quality of the harvest.

The two primary categories of mycorrhizae are ectomycorrhizal, and endomycorrhizal. Ectomycorrhizae are grown on hardwood and conifer trees, so is not of much interest to the cannabis gardener. Endomycorrhizae are grown with vegetables and cannabis roots. When inoculating your plant media with mycorrhizae, make sure to use the correct type.

Mycorrhizal inoculants are best applied at the beginning of the season. The goal is to establish and maintain a healthy colony. Once established, the colony should be self-perpetuating. A second application shortly before flowering is a common way to hedge the bet. Since cannabis requires more phosphorus in flower than in growth, and mycorrhizae assist in phosphorus availability and uptake, it is especially advantageous in flowering, particularly when combined with a carbohydrate additive. Mycorrhizae require time to develop, so adding new spores in late flowering adds little benefit.

Pythium oligandrum (Indoors and Outdoors) For damping off, gray mold, and Verticillium wilt.

Registered but not readily available in the United States, this cousin of the pathogenic fungi *Pythium ultimum* and *Pythium aphanidermatum*, *Pythium oligandrum* viciously attacks the hyphae of pathogenic fungi. In addition, it also secretes metabolites that inhibit the growth of other undesirable organisms. Developed in Slovakia, this product is presently supplied as either a powder, or as oospores encased in granules.

Trichoderma

Trichoderma is a genus of parasitic predatory fungi that develops a symbiotic relationships with roots. By wrapping itself around the roots it presents a physical barrier to pathogens. The fungi provide the roots with needed nutrients and use the root's exudates for nourishment. This increases plant resistance and vigor. The fungi also attack pathogens in the environment and use them for nourishment, too.

Trichoderma fungi usually have no sexual reproduction. However, the cells often contain several nuclei that have different numbers and sizes of chromosomes. They trade genetic material, which gives the organism the ability to evolve quickly to its environment. *Trichoderma* fungi are part of nearly all soil ecologies, where they play a prominent roll in keeping pathogens in balance.

There are several species of *Trichoderma* that are available commercially.

Trichoderma harzianum (Indoors and Outdoors) For damping off, gray mold, root rot, and Verticillium wilt.

Trichoderma harzianum is viable in a wide range of soil conditions and temperatures.

T. harzianum strains release chitinases, which are enzymes that destroy chitin. The pathogen cell walls are made from chitin so they dissolve as a result of the chemical onslaught. Its chemistry doesn't affect insects or nematodes. Brands include MycoGrow, Bountea Root Web, Root-Shield, and PlantShield.

Verticillium lecanii (Indoors and Outdoors) For aphids, spider mites, thrips, and whiteflies.

This cosmopolitan fungus is best used as a short-term agent to knock down high populations of insects until other biological controls can take over. Marketed by Koppert as Vertalec for aphids and Mycotal for whiteflies as well as for thrips and spider mites, *V. lecanii* does best in environments with a humidity level of 80 percent or higher for 10 to 12 hours per day, so it is excellent for clone rooms.

V. lecanii spores germinate and penetrate the insect's cuticle. After four to 14 days, the host

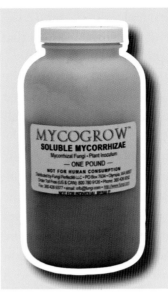

MycoGrow Soluble is a powdered mycorrhizal product containing many spores and species, plus other beneficial organisms such as *Trichoderma harzianum* and *Bacillus subtillus*.

dies from infection. Dead insects sprout a white fluff of conidiophores bearing slimy conidia that adhere to the bodies of passing insects.

Because of its high humidity requirements, do not spray *V. lecanii* when plants are flowering: The high humidity renders them vulnerable to gray mold. *Encarsia formosa* is slightly susceptible to *V. lecanii* at higher humidity levels but it is still a good companion control for whiteflies with this fungus.

Gall Midges. See Midges, Gall.

Geocoris pallens. See Big-eyed Bugs.

Geocoris punctipes. See Big-eyed Bugs.

Green Lacewings. See Lacewings.

Heterorhabditis bacteriophora. See Nematodes: *Heterorhabditis bacteriophora*.

Hypoaspis miles. See Mites, Predatory: *Hypoaspis miles*.

LACEWINGS

Lacewings (Aphid Lions) (Outdoors)

For aphids, caterpillars, leafhoppers, spider mites, thrips, and immature whiteflies.

Lacewings are found all over North America, Europe, and other continents and are noticeable in the garden and in natural areas. There are two types, green and brown, which are "second cousins." They feed on a combination of pollen, insect honeydew, other sugars, and insects. Some adults are more predatory, while others are more inclined towards a vegetarian diet.

Green lacewings are neon green and about ½ to ¾ inch long at maturity. They have two pair of glistening, transparent delicate wings that cover the length of the body, large golden eyes, and long antennae. The way they flutter in the air makes it seem like they aren't good fliers, but they have gotten around for millions of years, so the wings must be sufficient for their purposes.

Green lacewing.

Brown lacewings have dull-colored tan-to-brown wings that are translucent but not iridescent.

Adults are not big meat eaters. They live on honeydew and flower nectar, and in the greenhouse they are often fed a synthetic chow called Wheast. You can make your own Wheast by mixing one part sugar and one part yeast into a thin paste and applying it to a few leaves or spreading it on a plate.

Green and brown lacewings comprise two suborders, *Chrysopidae* and *Hemerobiidae*, which have different structures and reproductive strategies. There are hundreds of species in each suborder. The two suborders have the same life cycle and lifestyle but differ in internal layout a bit and in color. Many are non-differentiable by humans. Males of each type produce a particular low-frequency "song" that only attracts its own kind by vibrating their abdomen

Brown lacewing.

Aside from color, the main apparent difference between *Chrysopidae* and *Hemerobiidae*, is that the former, the green lacewing, places its egg at the end of a thin stalk that rises from the leaf surface, while *Hemerobiidae*, the brown lacewing, lays its eggs directly on the leaves.

The lacewing's larvae are the true beneficial, because they are general feeders. Species differ in shape from sleek to having humps but they usually look like a miniature version of an alligator. They sometimes have bristles along their bodies that collect debris, perhaps as camouflage. Instead of jaws, their mouths consist of hollow, sickle-shaped maxillae—straws that stick out in front of their bodies.

A larva walks around and when it touches a

living object such as an aphid, egg, or caterpillar, it grasps it with its front legs and injects it with protein-liquefying digestive venom. Then it sips the melted flesh, leaving an empty carcass.

Aphid lions are lacewing nymphs: their hollow jaws allow them to hold and suck juices from their prey.

Aphid lions feed on aphids, beetle larvae, caterpillar eggs, small caterpillars, cutworms, leafhoppers, mites, moths, thrips, and whiteflies. Green lacewings are whiteflies' biggest predator. They eat organisms larger than they are because they don't swallow, they just ingest the juices.

Lacewings vary greatly in their consumption depending on type and density of pests. An older larva can eat 30 aphids a day. Even after they are full, they sometimes attack aphids, grabbing them with their unique jaws, injecting them with venom, and tossing them off the leaves.

Aphid lions do not differentiate between beneficial insects and pests, and no one has been able to train them. Since they eat other beneficials don't use them in conjunction with other predators. They can be used in conjunction with parasitoids. Aphid lions are cannibalistic when other food sources are unavailable so the larvae should be introduced to the space individually, far away from each other.

After undergoing three molts the larva spins a cocoon and pupates inside it. The adult emerges in one to two weeks depending on the temperature. Over its monthlong existence as an adult, a female lays 200 eggs.

Both adults and larvae are nocturnal so they are not seen much in the daytime garden. Adults are often seen at night because they are drawn to light. These adults can be captured and placed in a greenhouse. If the introduction is successful, they will lay eggs, resulting in thousands of hungry aphid lions scouring the plant leaves in search of prey.

Outdoors, green lacewings are probably present in your area and you can attract them to your garden and keep any that are around from flying away. Supplying them with a no-fuss fast-food nectar substitute is a big incentive for the vegetarian adult lacewings to hang out. Indoors, the attraction to light may be a downside. Adults make suicide flights into the lights. For this reason they should be removed from the grow room or at least protected from the hot lamp by using a glass shield. These are commonly found on air- and water-cooled light reflectors.

Commercially, the larvae come individually packaged to protect them from each other because of their penchant for cannibalism. Nurseries sell both eggs and larvae. You are more likely to have success with the larvae rather than the eggs, because once a cannibalistic larva hatches, it may make the other eggs its first snack.

Green lacewing eggs.

Some populations of green lacewings are programmed for hibernation based on photoperiod. When the nights get long, the insects prepare for the cold season by going into a deep sleep where they wait for spring's longer days to re-activate them. In areas with cool or cold winters, expect populations to go into diapause. In areas without a clear definition of winter, such as Florida, the Gulf, and southern California, lacewings are active throughout the year.

Some nurseries sell populations that have been adapted to function under short-day light regimes. These are the best ones to use in a greenhouse or indoor garden. Two species, *Chrysopa camea* and *Chrysopa rufilabris*, are reared commercially for use in greenhouses. *C. camea* is the species most commonly offered by insectaries. However, it has some negative traits. It has strong migratory urges and will fly in air currents for three to four hours before it settles down. This behavior is partially mitigated in enclosed areas such as greenhouses, but it still exists. *C. rufilabris* doesn't have the migratory urges of *C. camea* and is a better candidate for introduction into enclosed areas. As long as it has an adequate supply of aphid honeydew, plant nectar, Wheast, or another pollen-honeydew substitute, it is content to hang out in the greenhouse or grow room.

Lady Beetles. See Beetles: Lady Beetles.

Ladybugs. See Beetles: *Cryptolaemus montrouzieri,* and Beetles: Lady Beetles.

Marcolophus caligenosus. See Mirid Bugs.

MIDGES, GALL

Gall midges, sometimes called gall gnats, are Ce-

cidomyiidae flies, which are predators of many crop pests. There are thousands of different species. Over 1,000 live in North America. Here are a few available for pest management.

Aphid Gall Midges (*Aphidoletes aphidomyza*) (Indoors and Outdoors)
For aphids.

Aphid gall midge.

The aphid gall midge is a Cecidomyiidae fly whose larvae are effective predators of aphids. They are fairly unselective; they eat most aphid varieties. Aside from eating aphids, midge larvae seem to enjoy attacking and killing aphids, so they are responsible for many more dead than eaten. When aphids are abundant, midges kill two to three times as many as they eat.

Predatory midge larvae in an aphid colony.

Aphid gall midges are 2 to 3 millimeters long and look a lot like a mosquito except for their long dangling legs and long antennae. They lay tiny bright orange eggs singly and in clutches, depending on the species, which hang from leaves by a thin, silk thread. They hatch in about three days. The bright orange, maggot-looking larvae eat between eight and 80 aphids to complete their three instars, or insect molting cycles. They accomplish this by attacking the aphid's back leg at the joint with their very powerful jaws. The midge paralyzes it using a nerve agent and then injects digestive enzymes into the prey. The flesh liquefies and the midge sucks it out, leaving an empty shell.

After growing for five to seven days, the larvae drop to the soil, dig themselves in a bit, and pupate. For this reason they must have access to planting media. They are not compatible with beneficial nematodes, which eat them, or in hydroponic gardens without planting mediums. They emerge one to two weeks later as vegetarian adults that eat honeydew. The females lay about 70 eggs over their two-week lifetimes.

Use one to two midges per 10 square feet. The adults are nocturnal and are rarely seen. They probably require a day/night cycle so gardens with continuous lighting do not provide them with the conditions they require. Provide a minimum of three hours of darkness daily. Midges go into diapause with a 12-hour darkness lighting regimen so should only be used in the vegetative room

Feltiella acarisuga (Indoors and Outdoors) For spider mites.

Feltiella acarisuga is found worldwide except for the tropics. It is a natural predator of mites. The midges are active as long as the environment remains moist and warm, between 60 to 70 °F.

The cycle begins when a female lays a single egg on a plant, close to a colony of mites. The oblong, translucent egg, 1 to .25 millimeters long, hatches two days after being laid. The larvae feed on all stages of mites for about five days, then spin a cocoon and pupate for another five days. Adults are about 2 millimeters long, live for about five days and are vegetarian, living on water and nectar. Wheast or sugar water may help them with their diet. They are very sensitive to pesticides.

MITES, PREDATORY

There are many species of predatory mites. Some mites live in the soil attacking root pests, and others on the foliage. The theory behind their use is that predators develop an equilibrium with the two-spotted mites so there are always a few prey and predators hanging around, causing little damage. Predator presence prevents pest population from peaking.

Apply predators more heavily when pest densities are high, and fewer predators when they are less dense. If food is scarce where the mites are located, they migrate. While predatory mites can be introduced in high concentrations to control an epidemic, the best strategy is to lower the pest population using other controls first. This reduces stress on the plants. Then add the predators to control the residual population.

The key is matching the predators with your specific environment. Each species has its own temperature, relative humidity, and light needs. I have never had success using predatory mites to control spider mites. Once introduced to the garden, they disappeared without a trace, with no apparent effect on the pest mite population. Although some people report success, most gardeners are disappointed.

Predatory mites are living organisms. To maximize their efficacy use them as soon as they are obtained. Place them directly on plant leaves no matter what carrier they are shipped on (usually bean leaves or vermiculite).

Each predator species has its own environmental requirements and preferred diet. Mites are listed for either indoors, outdoors, or both situations. These are general statements. Since greenhouses have controlled temperature and humidity range, they are considered indoor situations. Some of the mites that are listed for indoors because of high temperature requirements may be able to be used outdoors in sub-tropical and tropical areas, and even in warm temperate areas during the summer.

Amblyseius californicus. See *Neoseiulus californicus*

Galendromus occidentalis (Indoors)
For spider mites.

Galendromus occidentalis, the western predatory mite, is a good choice for hot dry conditions, so it is unsuitable for most indoor gardens and greenhouses. *G. occidentalis* prefers high temperatures of 80 to 110 °F and humidity of less than 30 percent, a lower humidity than most mites.

G. occidentalis doesn't have as voracious an appetite as other predatory mites. It eats only one or two spider mites or about five mite eggs daily. If its preferred food, spider mites, aren't available it is satisfied with red mites, pollen, fungus, and honeydew. Another nice characteristic of this species is its ability to walk on the webs that mites construct when the population reaches a critical mass.

Because of their small appetite and their ability to eat a variety of foods, this species can be kept alive in the garden ready to attack mite invaders. Invasion is likely to include only a few individuals, so a small population of predators is enough to eliminate them.

With new mothers starting to lay eggs every two weeks or so, and twice or three times as many females as males, *G. occidentalis* generates large populations very quickly when enough food is available. Commercial vegetable growers throughout the world use *G. occidentalis* preemptively on a regular basis intending that the predator greets immigrant mites as soon as they arrive.

Hypoaspis miles (Indoors and Outdoors)
For fungus gnats and thrips.

This native northern United States soil mite species does its work at night. Its preferred food is fungus gnat larvae, but it also feeds on springtails, thrips pupae, and other small soil insects. Tan in color and less than 1 millimeter in length, this tiny mite does best in moist soil and temperatures between 68 to 80 °F. Soil mites live in the top ½ inch of soil and cannot survive at deeper levels. They cannot tolerate standing water. Adults and nymphs consume up to five larvae per day. When prey is scarce, *H. miles* subsist on a diet of algae and plant debris.

H. miles is packaged commercially with adults in a pasteurized mix of peat and bran in quart shaker bottles. Each shaker bottle should contain 15,000 to 20,000 adults and may also contain another benign mite as a food source. Those containers arrive without food so the mites should be released as soon a possible. In packages that include the food, the mite can be kept up to two weeks in a cool, dry place.

H. miles is most effective when applied before the fungus gnat population reaches epic proportions. This is where the yellow sticky cards come in handy (see Sticky Cards in Section 1). When these are collecting less than 20 trapped adults per week, *H. miles* is an effective treatment. If the fungus gnat infestation is heavier, *H. miles* should be combined with other solutions.

These soil mites cohabit well with other beneficials such as *Steinernema* species. See Nematodes. Since *H. miles* exists in the soil substrate, it is unaffected by topical applications if a barrier is placed between foliage and soil, so it can be combined with foliar pesticide sprays. Place a cover over the planting medium so pesticides don't drip into it. *H. miles* is killed by most pesticides.

Iphiselus (Amblyselus) degenerans
(*Indoors*) For aphids, caterpillars, spider mites, thrips, and whitefly eggs.

Iphiseius degenerans is a predatory mite that attacks the first larval stage of *T. tabaci* and *F. occidentalis* but it eats mites and pollen as secondary foods. It prefers warm temperatures of 70 to 90 °F and humidity of 55 to 75 percent.

Originating from African and Eurasian areas of low latitudes where seasonal light shifts are slight, the species remains active in short-day lighting environments such as flowering rooms.

I. degenerans adults are dark brown and oval, about 0.7 millimeters long and mobile. Nymphs have a brown "X" on their backs. Eggs are laid along the veins on the undersides of leaves. They are transparent at first, but turn darker before hatching. Eggs hatch into adults in about 10 days. The adults live for about a month. Adults consume four or five thrips per day and lay one to two eggs per day.

I. degenerans attacks *N. cucumeris* but is compatible with minute pirate bugs. It doesn't thrive if there is a lack of pollen, which is a major part of their diet. Wheast may be an adequate pollen substitute.

I. degenerans tolerates neem oil and fungicides, but is sensitive to insecticides, especially soap and sulfur.

Mesoselulus longipes (Indoors and Outdoors) For spider mites.

Mesoseiulus longipes tolerates a wider humidity and temperature range than *P. persimilis*. Its eggs survive in humidity as low as 45 percent at 70 °F, but they need higher humidity as the temperature increases. The species survives a temperature range from 55 to 90 °F. It is not as aggressive a predator as *P. persimilis*. It eats eight to 12 immature spider mites or mite eggs a day, but it is tenacious. It can survive with little food for extended time periods and has a short lifespan, less than three weeks as an adult, so it reproduces quickly, resulting in large population surges when food is plentiful.

Neoselulus (Amblyselus) barkeri (Indoors and Outdoors) For spider mites and thrips.

A versatile mite, *N. barkeri* prefers temperatures of 66 to 90 °F and relative humidity of 65 to 72 percent. It preys on *T. tabaci*, *F. occidentalis*, and spider mites. It enters diapause under short photoperiods. Adults are 0.5 millimeters long and a deep tan color. *N. barkeri* reproduces very quickly. Adults can consume two to three thrips per day or 85 thrips in their lifetime. Females lay about one to two eggs a day.

N. barkeri works best on vegetative plants as it is compatible with other predatory mites and somewhat resistant to pesticides, but is not effective as *N. cucumeris*.

Neoselulus (Amblyseius) californicus

(Indoors and Outdoors) For aphids, leafhoppers, mealybugs, spider mites, and thrips.

This species thrives in a wide temperature range of 60 to 100 °F and in humidity of 50 percent or higher. It reaches sexual maturity in five to 10 days, then lays more than three eggs a day for three weeks. Its population is two thirds female. It eats mites, but also munches on other arthropods when they are scarce. *N. californicus* is not a thorough feeder. It leaves mites uneaten and moves to new leaves. This allows the host population to survive to keep the predator fed. This mite is sometimes identified as *Amblyseius californicus*.

Neoselulus (Amblyseius) cucumeris

(Indoors and Outdoors) For spider mites and thrips.

Amblyseius cucumeris.

This predatory mite feeds on thrips' eggs and the thrips' first instar, and also eats mites and pollen.

Which Mite Is Right?	
Pest	**Predatory Mite**
Aphids	*Iphiseius (Amblyseius) degenerans*
	Neoseiulus (Amblyseius) californicus
	Phytoseiid mites
Caterpillars	*Iphiseius (Amblyseius) degenerans*
Fungus gnats	*Hypoaspis miles*
Leafhoppers	*Neoseiulus (Amblyseius) californicus*
Mealybugs	*Neoseiulus (Amblyseius) californicus*
Spider mites	*Galendromus occidentalis*
	Iphiseius (Amblyseius) degenerans
	Mesoseiulus longipes
	Neoseiulus (Amblyseius) barkeri
	Neoseiulus (Amblyseius) californicus
	Neoseiulus (Amblyseius) cucumeris
	Neoseiulus fallacis (Amblyseius fallacis)
	Phytoseiid mites
	Phytoseiulus persimilis
Thrips	*Hypoaspis miles*
	Iphiseius (Amblyseius) degenerans
	Neoseiulus (Amblyseius) barkeri
	Neoseiulus (Amblyseius) californicus
	Neoseiulus (Amblyseius) cucumeris
	Phytoseiid mites
Whitefly eggs	*Iphiseius (Amblyseius) degenerans*

It prefers temperatures in the 66 to 81°F range, with humidity around 75 percent, and is effective on several species of thrips. In the absence of thrips, it eats mites and pollen.

N. cucumeris adults are 0.5 millimeters long

and beige to pink in color. At 77 °F they take six to nine days to develop from egg to adult. The adults eat about three to five thrips a day during their 30 day lifetime, during which they lay two to three eggs daily. This predatory mite competes with other mite predators and is cannibalistic. However, it is compatible with minute pirate bugs.

This mite is a good candidate for vegetative rooms because it prefers foliage rather than flowers, does best in humid conditions, and has a tendency to diapause under short-day regimens used for flowering.

This mite should be used in conjunction with other biological protectors because it only provides partial protection.

Neoseiulus fallacis (Amblyseius fallacis) (Indoors) Spider mites.

N. fallacis is a good candidate for grow rooms and greenhouses. It thrives over a wide temperature range into the mid 80s °F and successfully breeds at temperatures between 50 and 80 °F.

Females overwinter in crevices of tree bark if prey is available in the fall. They emerge in spring as soon as prey is available, but in reduced numbers because of heavy winter mortality. *N. fallacis* increases in number rapidly and adults become numerous by July. They live about 30 days as adults and lay an average of 40 to 60 eggs. Eggs are laid along the ribs of the undersides of leaves. Four to six generations are completed in a season in New York state. *N. fallacis* moves vigorously over plant surfaces in search of prey.

They do best at 50 to 80 °F with humidity between 60 to 90 percent and need pollen or Wheast to reproduce. They are very adaptable

Phytoseiid mites (Indoors and Outdoors)
For aphids, spider mites, and thrips.

These generalist predators are 0.5 to 0.8 millimeters long and can live in the soil or in leaf litter. They feed with a needle-like mouth structure called a chelicerae which allows them to pierce a mite and suck out the juices. They feed on honeydew and pollen if prey is minimal.

They have short life cycles, about one week, high reproductivity (40 to 60 offspring per female), and eat about 20 prey per day. They are particularly fond of spider mites, but will eat thrips, aphids, and two-spotted spider mites.

Phytoseiulus persimilis (Indoors)
For spider mites.

The yellow/orange predatory mite is on a leaf with a mature and immature spider mite: *Phytoseiulus persimilis.*

P. persimilis is a voracious spider mite-eater. It consumes 10 to 20 mites and mite eggs per day. Females lay 30 to 60 eggs over their 50-day lifespan. To reproduce successfully it needs high humidity. At 50 percent relative humidity, less than 10 percent of the eggs hatch. At 70 percent relative humidity, more than 90 percent survive. They do best at temperatures of 70 to 85 °F and tolerate humidity between 60 to 90 percent.

P. persimilis is a thorough eater and hunts the host population to extinction. Since its diet is based solely on mites, this leads to its extinction as well. Then it has to be re-introduced when there is a new infestation. It attacks some other predatory mites such as *N. cucumeris*.

Nabis alternatus. See Damsel Bugs.

NEMATODES, BENEFICIAL
Heterorhabditis and **Steinernema**
(Indoors and outdoors in temperate climates)
For caterpillars, fungus gnats, root aphids, soil-dwelling beetles, and thrips.

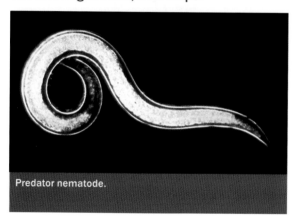
Predator nematode.

There are thousands of types of nematodes that occupy every environment and fill many ecological niches. The types we are interested in are entomopathogenic, or insecticidal worms, in the genera *Steinernema* and *Heterorhabditis*, which are soil inhabitants. Nematodes are found in virtually all natural soils.

All of these species move through the media following their prey's scent trails. When they come in contact with an insect, they find an opening between its chitin cuticles—a wound, or in the case of the *Heterorhabditis*, they use their single sharp tooth to pierce the insect.

The nematodes inject a bacterium into the opening that paralyzes its host by digesting the tissue. At the same time the bacterium produces antibiotics and chemicals that repel other nematodes. The victim dies within 48 hours. The nematodes invade and slurp the liquefied tissue-bacteria mixture. After feasting, the nematodes have sex and lay eggs that hatch inside the host. Several generations of nematodes feast on big game. When food becomes scarce the nematodes produce a traveling generation, specially adapted to deal with the outside world. These emerge from the cadaver. The juveniles eat the remaining food, then head out in search of more prey to restart the cycle.

Beneficial nematodes parasitize a variety of insects that spend at least part of their lives in or on the soil.

Entomopathogenic nematodes die from dry conditions, exposure to UV light such as sunlight, and storage in hot conditions. They are not sensitive to most pesticides. They are most active when the soil or media is between 68 to 85 °F.

Dead insects turn brown or tan when infected by steinernematid species, and red and gelatinous when infected by heterorhabditids, and the tissue develops a gummy consistency. The symbiotic bacteria heterorhabditids use is slightly luminescent so the cadavers glow. This is a foolproof diagnostic for the genus although the symbiotic bacteria provide the luminescence. Rotting, black cadavers associated with putrefaction indicate that the host was not killed by entomopathogenic nematodes.

The nematodes are sold as third stage juveniles, ready to search, find, and infect pests. Depending on the carrier they are held in, they are either watered into the soil or placed on top of it.

Watering from the top of containers flushes nematodes from the top of the soil, making the treatment ineffective. In a test the worms were not present in different potting soils after 30 days of overhead watering. With sub-irrigation, watering where pots are placed in water-holding trays, nematodes were detected 120 days after application.

Many species of beneficial nematodes are available. The ones you are most likely to use are *H. bacteriophora*, *S. carpocapsae*, and *S. feltae*.

Heterorhabditis bacteriophora (Indoors and Outdoors) For beetles, fire ants, and thrips.

H. bacteriophora is a cruiser; third-stage larvae track down hosts by following their CO_2 and feces. They enter through openings such as mouth, anus, or spiracle, or by piercing the host using a sharp tooth. Once inside, the nematode releases an enzyme that inhibits the host's immune system. Then, they release symbiotic anaerobic bacteria (*Photorhabdus luminescens*), which actually do the killing within 48 hours.

H. bacteriophora has superior host-penetrating capabilities over *Steinernema* species and live deeper in the soil. However, *Steinernema* species store better over longer periods of time.

Sometimes infestations diagnosed as root aphids are actually a type of soil-dwelling beetle. *H. bacteriophora* is effective against these pests. Infective juveniles remain effective for only a few days, so must be reapplied fairly often. To be effective they need a minimum temperature of 68 °F.

In some species of hosts, the parasitized cadavers glow faintly in the dark.

H. bacteriophora is available commercially under a number of names: Nemaseek, Heteromask, and Terranem.

Steinernema carpocapsae (Indoors and Outdoors) For catepillars, fungus gnats, and some larvae.

S. carpocapsae dwells near the soil surface, about 5 centimeters deep. It is most effective against various caterpillars and to a lesser extent, fungus gnats. It is an ambush forager. It nictates, that is, stands upright on its tail near or at the soil surface. It attaches itself to passing hosts and enters through one of the spiracles (insect breathing holes), which the nematode finds by following a trail of CO_2.

Phasmarhabditis hermaphrodita (Outdoors) For slugs and snails.

This parasitic nematode lives in the soil and actively seeks out slugs and snails in the soil. Preferring slugs of the genus *Deroceras* and *Arion*, and the snail *Helix aspersa* as well as some others, this tiny roundworm is about 1 millimeter long. Within three to five days of infection by this nematode, slugs and snails sicken and stop feeding. Death usually follows within 10 days.

Juveniles arrive in packets and can only be stored for one or two days in a cool, about 35 °F, dark place. They should be poured or sprayed onto moist soil at night. The soil must be kept moist for two weeks after application to supply the nematodes the water they need to move through the soil.

Supplied commercially as Nemaslug.

Steinernema feltiae (Indoors and Outdoors) For fungus gnats, fruit flies and other *Dipteras*, and thrips.

S. feltiae is the best nematode to use for fungus gnats, fruit flies and other *Dipteras*, thrips, and other insects. It is also effective against root aphids. It tolerates cold better than other nematodes.

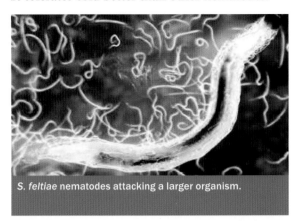
S. feltiae nematodes attacking a larger organism.

It frequents the top of the soil to a depth of 7.5 centimeters. It uses both a cruiser technique when hunting prey, moving through the soil on a thin layer of water held between soil particles, as well as ambush foraging. When it finds a host, the nematode invades its body through orifices, such as the mouth, spiracles, or anus, or else burrows in through a thin place in the cuticle.

Once inside, *S. feltiae* releases a symbiotic bacterium, *Xenorhabdus sp.*, which multiplies quickly causing rapid death and liquefaction of the bodily tissues that the nematodes feast on.

Apply in the evening or on cool, cloudy days to allow the nematodes time to burrow into the soil and avoid desiccation as the sun and heat of the day begin. Reapply at six-week intervals. *S. feltiae* brands include NemAttack, NemaShield, Scanmask, and Entonem.

ANTIdote contains three types of parasitic nematodes that naturally attack ants. Although it is advertised for fire ants, the manufacturer's spokesperson said it may be effective against other kinds of ants, including the Argentine ant.

PROTOZOA
Nosema locustae (Outdoors)
For grasshoppers.

This microsporidian (spore-forming) protozoan is a naturally occurring parasite of grasshoppers found throughout the natural population. The protozoa infect the fat cells and other internal organs of grasshoppers and then produce large numbers of spores within the insect. Infected individuals die from declining health causing them to eat and breed less and to become so weak they fall prey to predators.

N. locustae is not a quick fix for a grasshopper problem because it takes a long time for the organism to become established and do its damage. However, with patience, *N. locustae* is very effective in reducing grasshopper populations. Eggs laid by infected grasshoppers are also infected and the subsequent nymphs pass the protozoa along to their offspring. However, due to the debilitating nature of the protozoa, infected grasshoppers breed with less success, slowing the spread of the organism.

N. locustae is available commercially, but it is used primarily for large acreages with huge grasshopper problems, usually on rangeland. It is not effective for backyard use.

Orius species. See Minute Pirate Bugs.

Rove beetle. See *Atheta coriaria*.

Seaweed. See Section 2: Seaweed.

SNAILS, PREDATORY
Rumina decollate (Outdoors)
For slugs and snails.

Decollate snails are predators of snails and slugs.

They are used commercially in the citrus groves of southern California to prey on snail eggs and immature garden snails. Slow to establish an adequate population, *R. decollate* takes one to two years to control a large snail population. They are supplied as dormant adults that can be stored up to a week in a cool 46 to 50 °F dark place. Release one to 10 snails per square meter into moist, shady areas.

Their use is restricted in some places because of the threat they pose to native populations of endangered mollusks. They have naturalized in the Sun Belt of the United States from Florida to California and are considered an invasive species in some states due to their impact on native species. In California they cannot be released outside of the southernmost counties and are not allowed in the north. Should decollate snails wipe out their prey population, they feed on seedlings and other vegetation.

Ryzobius lophanthae. See Beetles: Lady beetles.

WASPS, PARASITIC

Aphidius colemani wasps (Indoors and Outdoors) For aphids.

A small parasitic wasp, native to North America that parasitizes the green peach aphid, black bean aphid, and cotton aphid, *A. colemani* occurs naturally outdoors. It is a good searcher, and seeks new aphid colonies when aphid populations decline.

A. colemani's life starts when its mother deposits a single egg in a non-parasitized aphid.

The egg hatches and the larva grows inside the aphid. In 10 days at 77 °F, to two weeks at 70 °F, the mature 3-millimeter-long wasp emerges from the mummified aphid. It feeds on honeydew it finds on leaves and, if female (about two-thirds of the population), it begins laying eggs in aphids, which it searches for by detecting the aphid's alarm chemical, (E)-beta-farnesene.

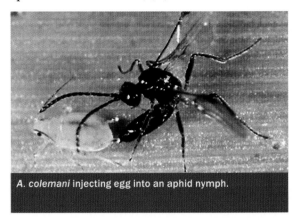

A. colemani injecting egg into an aphid nymph.

Before *A. colemani* mates, the eggs produce males. After fertilization, females result. Over her two-week lifespan, she will lay between 10 and 20 eggs a day, parasitizing a total of 150 to 300 aphids. They survive over a range of conditions but perform best at 65 to 85 °F with high humidity.

A. colemani is usually shipped as parasitized aphid mummies from which adults will emerge. Application rates call for 1,000 to 3,000 per acre and one per tomato or pepper plant in an uninfected greenhouse. In infected greenhouses, the recommended release rate is about one to five per square foot. They are usually sold 500 to 1,000 per package so their release in a small greenhouse or indoor garden results in fast action. These insects can fly away so they aren't recommended for release in small garden areas.

Aphidius matricariae wasps (Indoors and Outdoors) For aphids.

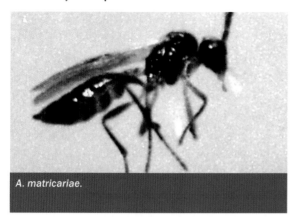

A. matricariae.

Aphidius matricariae is a North American native wasp. It follows the description of *A. colemani* with the exception that it does not attack cotton aphids. Other wasp parasites are available as well. *Aphidius ervi* is used primarily for control of the potato aphid and *Aphidius abdominalis* can be used to control green peach aphids but prefers potato aphids.

Ceranisus menes wasps (Indoors) For thrips.

This wasp parasitizes two types of thrips: the western flower thrips (*F. occidentalis*), and onion thrips (*T. tabaci*). It is found worldwide. Females insert their ovipositors into the bodies of thrips nymphs. Once the nymph is paralyzed by the female's sting, a single egg is laid inside the body. The eggs hatch into larvae, which eventually kill the host. *C. menes* larvae then exit the body *Alien* style and drop into the soil to pupate. *C. menes* also kills adult thrips by wounding them with their ovipositor, then drinking the fluids that ooze from the wounds.

Encarsia formosa wasps (Indoors and Outdoors) For whiteflies.

Encarsia formosa.

The most successful biocontrol organism for the control of whiteflies in greenhouses is a tiny golden parasitic wasp called *Encarsia formosa*. Its preferred host is the greenhouse whitefly. A particular race, Nile Delta, attacks all major whitefly species so identifying the whitefly strain is not necessary. Adult wasps use sight and smell to find whiteflies. They lay eggs inside the whitefly larvae, preferably during their second instar. Whitefly larvae turn black as the wasp develops inside the body. After 15 to 25 days, the adults wasps emerge.

Encarsia formosa injecting eggs into an aphid.

E. formosa does better in environments with abundant light, high humidity, and temperatures of about 73 °F, making it best for indoor uses in most areas. It is most effective against the greenhouse whitefly and not as effective against the silverleaf or sweet potato whitefly. Three other parasitic wasp species, *Encarsia luteola*, *Eretmocerus californicus*, and *Eretmocerus eremicus* are commercially available for control of these species.

The pupae of *E. formosa* are available on cards, which are hung in the growing area. Adults hatch and wing their way to the waiting buffet. Whiteflies must already be present before introducing the wasps so the newly emerging adults have food readily available.

Metaphycus helvolus wasps (Indoors)
For mealybugs and scales.

Metaphycus helvolus is a tiny, parasitic wasp that was introduced into the United States from South Africa in 1937 to control black scales in California citrus orchards. *M. helvolus* preys on a wide range of hosts, including many scales like the hemispherical and the brown scale.

The adult *M. helvolus* wasp is 1 millimeter long. Males and females differ in color. The female is orange-yellow and the male is dark brown. Females deposit their eggs in second and third instar scales, which have a thin waxy coating. Male wasps die soon after mating but females live up to 90 days. Females feed on adult scales by drilling a hole in their exoskeleton using their ovipositors, and then sipping the liquid as it seeps up from the wound. Well-fed wasps can lay up to five eggs a day for months, so they lay an average of 400 eggs over their lifetime.

The larvae hatch in a couple of days and then spend their entire development period inside the scales. They emerge as adult wasps after two or three weeks by cutting a hole in the now-mummified host. The wasps have a great appetite but don't always finish their meals, so they leave many scales oozing to death while they drill and taste a new victim. The wasps also feed on the honeydew exuded by the scales. The wasp's entire life cycle is based on the scale hosts.

M. helvolus is an effective biocontrol agent of hemispherical scale in the greenhouse and grow room but it provides only partial control of brown soft scales because some of the wasp larvae are encapsulated and killed by the scales. Release *M. helvolus* evenly over the plants three times at 10-day intervals at a rate of five to 10 per infested plant, 10 per square meter. Two weeks after release, mummified pupae with emergence holes will be visible and the scale problem will diminish and disappear over two months. The wasps are sensitive to pesticides so make sure there are no residues present when the wasps are released.

Synacra pauperi (Indoors and Outdoors)
For fungus gnats.

This parasitic wasp lays its eggs inside fungus gnat larvae. These wasps have been used successfully in Swedish greenhouses. The University of Illinois was able to eliminate a chemical spray program for their Conservatory and Plant Collection by using *Synacra pauperi* exclusively for fungus gnat control. This wasp is not available commercially but is a natural parasitoid of fungus gnats and often follows their population.

Thripsobius semiluteus wasps (Indoors and Outdoors) For thrips.

Available in Holland, this tiny wasp parasitizes greenhouse thrips (*H. haemorrhoidalis*). It is native to Africa, Australia, Brazil, and South Africa. Optimal temperature is 73 °F with 50 percent relative humidity.

Adults prey on first and second instar nymphs and the parasitized hosts turn into black, swollen mummies. Female wasps lay about 40 eggs. Adults emerge from the parasitized insect cadaver to repeat the 22-day life cycle. Pupae come in vials that are attached to plants high in the canopy. Do not use insecticides within 30 days before release.

Trichogramma Wasps (Indoors and Outdoors) For caterpillars.

Trichogramma wasp laying eggs in moth egg.

Wasps in the *Trichogramma* genus are also known as chalcid wasps. These small parasitic members of the order Hymenoptera are effective against many species of Lepidoptera. They lay their own eggs inside moth or butterfly eggs, which hatch into larvae that feed off the contents of caterpillar eggs. The eggs hatch and go through growth and metamorphosis inside the Lepidoptera egg, emerging as adult wasps rather than caterpillars. They solve caterpillar problems without the introduction of any chemicals, natural or otherwise. As such, *Trichogramma* has become one of the most popular biocontrol measures by farmers and gardeners around the world.

These wasps are so tiny, about 0.5 millimeter from one tip of the wing to the other, that you may not even see them. Adults have pear-shaped wings with a single vein and fringed edges. The best use of *Trichogramma* is to release them when either moths or caterpillars are first sighted in the garden. Depending on the size and location, multiple weekly releases may be needed over a three to six week period. A good rule of thumb for each release is about one wasp per square foot.

Trichogramma is available commercially. Its pupae are glued to small paper or cardboard squares, which are placed in the garden. The wasps hatch and breed without any further effort on the part of the gardener.

SECTION 4.
Outdoor Strategies

Many of the barriers, pesticides, and fungicides previously described work as well outdoors as they do in an indoor garden or greenhouse. However, there are obvious differences to pest management when growing outdoors. In controlled indoor environments—which may be completely sealed from the outdoor environment—the objective is often "zero tolerance" to any external influences.

Outdoors, the farmer has to analyze problems and set specific goals. Realistically, the best strategies are often not the complete elimination of a pest; it is to deter the pest from having any interest or access to cannabis plants. This section talks about general outdoor strategies related to insects and mammals.

ANIMALS AS PEST CONTROL

The presence of other animals may keep some large and small pest animals at bay.

Birds: Attracting Wild Birds
(gophers, grasshoppers, rats)

Attracting wild birds to your growing area adds another layer of help. Larger species of birds such as robins, kestrels, grouse, and meadowlarks are capable of eating hundreds of grasshoppers in a day. During the nesting season, that number increases, as grasshoppers are a staple in the diet of many young birds.

Owls, hawks, and other natural predators lower the rat population a bit, but they will not eradicate an infestation. Owls prey on gophers, but they hunt over a very large area and usually don't significantly impact a particular population. However, making them local by putting up owl boxes to attract them helps control the local population. Owls also eat other pests including rats and mice.

Cats (rats)

Cats are great biological control for rats and mice. Cats love to roam and hunt outdoors, and their presence makes rats think twice about taking up residence there, creating a good form of biological control. Rodents have a built-in aversion to being near cat odors. When cats spread their scent around an area, mice and rats avoid it.

Chickens. See Fowl.

Dogs (deer, rats)

Deer and rats stay away from dogs. Dachshunds and some terriers are particularly good ratters and can be used to keep the garden free of rats as long as they are in the garden when rats are active. A dog in the garden is all deer need to remind them that they are deerona non grata.

Fowl (grasshoppers, slugs, snails)

Fowl are chickens, ducks, or geese. When allowed to run free in the garden they are eager insect, slug, and snail gobblers. While ducks and geese are efficient hunters, they can tramp down foliage and eat young seedlings. They are best used when the foliage is above their pecking height. Allowing chickens, turkeys, guinea hens, and other fowl into the garden area is an excellent control for a number of insects including grasshopper of all sizes. Turkeys are particularly fond of grasshoppers and will stubbornly pursue a single individual until it is caught and eaten. Fowl should only be used among tall, strong, sturdy plants because the birds will peck at the succulent greens and throw their weight around, knocking the plants down.

Frogs and Toads (slugs, snails)

Frogs and toads eat snails and slugs. Frogs are the better predator of the two. Slugs are a quarter of their diet. Attract frogs to the garden by providing a clean, reliable water source and places to hide such as flower pots turned sideways.

Opossums (slugs, snails)

Opossums are marsupials that eat small animals including insects, frogs, slugs, and snails. They prefer snails because they don't produce protective mucus. When opossums are around, their diets help keep down the slug and snail population.

Owls. See Birds.

AGRICULTURAL CONTROLS
Brassica Residues (fungi such as gray mold and root rot)

Crops of cabbage and *Brassica* species, such as cultivars of broccoli, winter rape (*Brassica napus*), mustard, and wild mustard produce glucosinolates, which decompose into nitriles and isothiocyanates in the soil. These compounds are effective against pathogenic fungi, and some insects.

Glucosinolates and isothiocyanates are thought to provide anti-carcinogenic properties to mammals.

Plant *Brassicas* in the fall after harvesting. The variety you use for cover cropping varies by your location. In late fall, before freezing, turn the crop under. The cell walls rupture releasing glucosinolates, which bacteria convert to nitriles and isothiocyanates.

The effect of these chemicals can be enhanced by covering the soil with a 1- to 2-millimeter-thick clear polyethylene tarp and securing the edges to increase soil solarization. The sun heats up the soil and the heat is retained under the tarp. The warm temperature increases the release of the *Brassica*'s protectant chemicals into the planting medium. See Soil Solarization below.

Cleared land (grasshoppers and thrips)

Cleared land leaves insects with no source of food or shelter so they are vulnerable to predation. Use a 10-foot dead zone around the greenhouse to eliminate areas where pests can live, breed, and swarm. Remove plant debris from the garden, the growing area, and from greenhouse benches to eliminate locations where thrips can hibernate.

Companion Plants (grasshoppers, gray mold, leafhoppers, slugs, thrips, and whiteflies)

Companion plants can serve either of two purposes in the garden—they either repel insects from the desired plant, or they attract the pests to them instead and are called "trap crops." Heavily scented plants such as mints repel leafhoppers and other insects, as do germaniums and members of the onion family, such as garlic and chives. Oregano, thyme, and rosemary produce resins and essences that repel or kill a wide range of pests, including leafhoppers. The presence of cilantro will discourage grasshoppers.

Nasturtiums (*Tropaeolum majus*), African and French marigolds (*Tagetes*), calendula, and Peruvian cherry or Cape gooseberry (*Physalis peruviana*) repel whiteflies. Shoofly, which is *Nicandra physalodes*, tobacco, and members of the nightshade family, on the other hand, attract whiteflies and capture the little beasts on sticky, hairy leaves. Cucumbers, eggplant, sweet potato, cotton, and tomatoes attract whiteflies. Attractants can be used as indicator crops, letting you know when an infestation is beginning.

A garlic plant for every 10 square feet of canopy repels thrips. This can be helpful when the marijuana plants are small. As they grow larger and the garlic is further away from the canopy, this becomes less effective. Using garlic juice spray every few days repels the pests. If a gardener is concerned about the devlopment of gray mold, planting *Brassicas* in the area can be helpful.

Plants as Indicators

Majestic, Blue Magic, and Calypso petunias are irresistible to thrips and are used to indicate their presence. These plants work as indicators because they are highly susceptible to tomato spotted wilt virus, which is spread by western flower thrips. Within three or four days of infection, a distinct brown halo appears at the feeding site on the leaves, followed by a lesion a few days later. The virus does not spread from the petunias to other plants in the area, as thrips prefer the petunia flowers. Symptoms do not appear on the flowers. To encourage the thrips to concentrate on feeding on the leaves, most of the flowers should be removed. Use trap crops with some caution, since they attract pests and clearly the goal is not to attract the pests to the garden.

ENVIRONMENTAL CONTROLS

Fences (for deer, humans, and other large animals.)

Fences are more a deterrent than actual prevention. A sufficiently motivated animal can generally defeat a common fence if it believes that the reward is worth the effort. Fences are best employed in situations where the motivation to breach is moderate to low. See Deer and Gopher Chapters.

Flooding. See Chapters on Gophers and Verticillium Wilt.

Insect Netting. see Section 1: Insect Netting.

Soil Solarization (ants, damping off, fungus gnats, grasshoppers, root aphids, root rot, snails, thrips, and Verticillium wilt.)

Soil solarization is a non-chemical method of eliminating weeds, pests, and diseases. Clear polythene sheeting is placed over the soil so it retains heat when sunlight shines on it. This creates a lethally hot environment by retaining the heat.

When soil is covered during the two hottest months, June and July—in some places extending to May and August or September—it suppresses bacteria, fungi, mold, weeds, weed seeds, insects, and other pests. This technique is normally used in areas with lots of sun and high temperatures during part of the year, but it can be modified for cooler gardens and climates as well.

In addition to suppressing pests and diseases, solarization changes the soil chemistry a bit, causing organic matter to release nutrients. This promotes an increased growth response from the additional nutrients available, and stimulates plant resistance to pests and disease.

First remove debris, and soak the soil so it is thoroughly moist. Place polythene sheeting over the area and secure the edges using rocks, soil piles, or lumber. Using a meat thermometer, measure the soil temperature 3 to 4 inches below the surface. When the temperature reaches 180 °F for 30 minutes, the soil is pasteurized and is free of weeds, pests, and most diseases.

In cooler areas soil may never reach 180 °F. When the temperature reaches 120 °F daily over a two-week period, the soil is considered pasteurized. Rather than killing pathogens, the high heat weakens them and makes them susceptible to attack by other organisms. One problem with the 180 °F treatment is that beneficial organisms, including beneficial nematodes and mycorrhizae, are killed along with the pathogens. The soil should be inoculated with mycorrhizae and other beneficials before planting.

Thinner plastic is more effective because more light energy passes through the plastic. On the other hand, when there are large temperature differences between day and night, sheets of bubble packing material help the soil retain heat longer at night.

Traps. See Chapters on Deer, Gophers, and Rats.

Traps are the best method of controlling mammals such as deer, gophers, and rats. See specific pests for details on how to trap them.

PARTIAL BIBLIOGRAPHY

Agboka K., Tounou A.K., Al-moaalem R., Poehling H-M., Raupach K., and Borgemeister C. "Life-table study of Anagrus atomus, an egg parasitoid of the green leafhopper Empoasca decipiens, at four different temperatures." *Bio-Control* 49, (June 2004): 261-275.

Agraquest. Serenade Max. http://www.agraquest.com/agrochemical/products/fungicides-serenade-max.php.

Agrios, George N. *Plant Pathology*, 5th edition. (New York: Academic Press, 2005).

Ahlstrom, Dr. Ken. Plant Industry Division: North Carolina Department of Agriculture. Personal communication.

Alderton, David. *Rodents of the World*. (Blandford Press, 1996).

Alexopoulos, Constantine John. *Introductory Mycology*. (New York: John Wiley & Sons, 1966).

"All About Hantaviruses," Center for Disease Control, http://www.cdc.gov/ncidod/diseases/hanta/hps/noframes/rodents.htm

Allotment Growing Diary Plus. "Vegetable, Fruit and Herb Growing on My Allotment: Controlling Whitefly with Parasitic Wasps Encarsia Formosa." http://www.allotment.org.uk/garden-diary/37/controlling-whitefly-with-parasitic-wasps-encarsia-formosa/.

Alston, Diane G. "Utah Pests Fact Sheet: Onion Thrips (Thrips tabaci)." http://extension.usu.edu/files/publications/factsheet/ENT-117-08PR.pdf.

Arbico Organics. "Leafhoppers and Planthoppers." http://www.arbico-organics.com/category/pest-solver-guide-leafhoppers-planthoppers.

Arbico Organics. "Live Ladybugs *Hippodamia convergens*." http://www.arbico-organics.com/product/Live-Ladybugs.

Arbico Organics. "Praying Mantis Egg Cases—*Tenodera aridifolia sinensis*." http://www.arbico-organics.com/product/Praying-Mantis-Egg-Cases/pest-solver-guide-leafhoppers-planthoppers.

Arnobrosi. "The Biology of Snails." http://arnobrosi.tripod.com/snails/evo.html.

Arnobrosi. "The Trail of the Snail." http://www.arnobrosi.com/snails/snail.html.

Backyard Nature. "Snails and Slugs." http://www.backyardnature.net/snail_sl.htm.

Bailey, Winston J., and Marsusi Nuhardiyati. "Copulation, the dynamics of sperm transfer and female refractoriness in the leafhopper Balclutha incisa (Hemiptera: Cicadellidae: Deltocephalinae)." *Physiological Entomology* 30, No. 4 (2005): 343-352. Accessed May 18, 2010.

Bambi Bucket Systems. http://www.sei-ind.com/products/bambi-bucket-systems

Batra, S.W.T. "Some insects associated with hemp or marijuana (Cannabis sativa L.) in Northern India." *Journal of the Kansas Entomological Society* 49, No. 3 (1976): 385.

Bell, Michelle L. and James R. Baker. "Comparison of Greenhouse Screening Materials for Excluding Whitefly (Homoptera: Aleyrodidae) and Thrips (Thysanoptera: Thripidae)." *Journal of Economic Entomology* 93, No. 3: 800–804.

Berry, Ralph E. "Western Damsel Bug." *Insects and Mites of Economic Importance in the Northwest* 2nd Ed. (1998).

Biobest Biological Systems. "PreFeRal® WG: A biological insecticide." http://www.biobest.be/v1/en/producten/biopesticiden/paecilomyces.htm

Boudabous, A., Sadfi- Zouaoui, N., Essghaier, B. Hajlaoui, M.R., Fardeau, M.L., Cayaol, J.L., and Olivier, B. 2008. "Ability of Moderately Halophilic Bacteria to Control Grey Mould Disease on Tomato Fruits." *Journal of Phytopathology* 156 (2008): 42–52.

"Botrytis cinerea." *Encyclopædia Britannica* (2010). Retrieved September 13, 2010, from Encyclopædia Britannica Online: http://www.britannica.com/EBchecked/topic/1460625/Botrytis-cinerea

Bradley, Fern Marshall, Barbara W. Ellis and Deborah L. Martin. *The Organic Gardener's Handbook of Natural Pest and Disease Control* (New York, NY: Rodale, 2009).

Brogran, Carlos E., Kevin M. Heinz. "House and Landscape Pests: Whiteflies." AgriLife Extension: Texas A & M System. http://docs.google.com/viewer?a=v&q=cache:Pi8DCgyT-RMJ:repository.tamu.edu/bitstream/handle/1969.1/87303/pdf_209.pdf%3Fsequence%3D1+damsel-+bugs+whitefly+control&hl=en&gl=us&pid=bl&srcid=ADGEEShlHhz-Tr6hi8XsFlE0kPOPYAC7EFQenG4_Xe80r-TUMjWckxY2mhxUHMNYsPfrja.

Buda, Elizabeth. "Bubonic Plague." NSF GK-12 Graduate Fellows Program. University of North Carolina at Wilmington. Award # DGE-0139171.

Burgess, D. R., T. Bretag, and P. J. Keane. "Biocontrol of seedborne Botrytis cinerea in chickpea with Clonostachys rosea." *Plant Pathology* 46 (1997): 298 – 305.

Burrows, M. "Energy storage and synchronization of hind leg movements during jumping in planthopper insects (Hemiptera, lssidae)." Journal of Experimental Biology February 2010; 213(3):469-478, Accessed May 17, 2010.

Burrows, Malcolm. "Anatomy of the hind legs and actions of their muscles during jumping in leafhopper insects." *Journal of Experimental Biology* 210, no. 20 (2007): 3590-3600.

Burrows, Malcolm. "Kinematics of jumping in leafhopper insects (Hemiptera, Auchenorrhyncha, Cicadellidae)," *Journal of Experimental Biology* 210, no. 20 (2007): 3579-3589.

Burrows, Malcolm. "Jumping performance of planthoppers (Hemiptera, lssi-

dae)," *Journal of Experimental Biology* 212, no. 17 (2009): 2844-2855.

Burton, D. W. "How to be Sluggish," *Tuatara* 25, Issue 2 (January 1982).

Brunt, A. C. "Plant Viruses Online: Descriptions and Lists from the VIDE Database. Version: 20th August 1996." http://biology.anu.edu.au/Groups/MES/vide/

Cabanillas, H. E. and J. R. Raulston. "Evaluation of Steinernema riobravis, *S. carpocapsae*, and Irrigation Timing for the Control of Corn Earworm, *Helicoverpa zea*." *Journal of Nematology* (1996 March): 75–82.

Caithness.org. "The Black Rat's Home in Roman Britain." http://www.caithness.org/caithnessfieldclub/bulletins/1984/october/black_rats_home_in_britain.htm

California Department of Agriculture. "Plant Quarantine Manual." http://pi.cdfa.ca.gov/pqm/manual/htm/106.htm

Campbell, Neil A., Jane B. Reece, Lawrence G. Mitchell, and Martha R. Taylor. "Biology: Concepts and Connections." (Benjamin Cummings: 2003).

Capinera, John L. "Encyclopedia of Entomology, Volume 4." (Springer: 2008).

Chase, Ronald, and Katrina C. Blanchard. "The snail's love-dart delivers mucus to increase paternity." Proceedings of the Royal Society of Biological Sciences. http://rspb.royalsocietypublishing.org/content/273/1593/1471.short

"Chemoreception." *Encyclopædia Britannica* 2010. http://www.britannica.com/EBchecked/topic/109023/chemoreception

Chiu-Chen Huang, Chen-Hsiung Yang, Yuko Watanabe, Yung-Kung Liao, and Hong-Kean Ooi. "Finding of *Neospora*

caninum in the wild brown rat (*Rattus norvegicus*)." *Veterinary Research* 35 (2004): 283–290.

Choquer, Mathias, Elisabeth Fournier, Caroline Kunz, Caroline Levis, Jean-Marc Pradier, Adeline Simon and Muriel Viaud. "*Botrytis* cinerea virulence factors: new insights into a necrotrophic and polyphageous pathogen," FEMS 277 (2007): 1–10.

Cirrus Image. RedBanded Leafhopper. http://www.cirrusimage.com/homoptera_leafhopper_Graphocephala_coccinea.htm

City of Carlsbad California. Plants that Repel Pests. http://www.carlsbadca.gov/services/departments/water/pages/gardening-tips.aspx

Cloyd, Raymond A. "Know Your Friends: The Entomopathogenic Fungus Metarhizium anisopliae." http://www.entomology.wisc.edu/mbcn/kyf607.html.

Conrad, Jim. "Snails & Slugs." http://www.backyardnature.net/snail_sl.htm.

Cornell University. "Biological Control: A Guide to Natural Enemies in North America. Phasmarhabditis hermaphrodita." http://www.nysaes.cornell.edu/ent/biocontrol/pathogens/phasmarhabditis_h.html

Cornell University. "Botrytis Blight." http://plantclinic.cornell.edu/FactSheets/Botrytis/Botrytis_blight.htm

Cote, Kenneth W., Edwin Lewis, and Eric Day. "Virginia Cooperative Extension: Whiteflies." http://pubs.ext.vt.edu/444/444-280/444-280.html.

Cranshaw, W. S. and R. Hammon. "Grasshopper Control in Gardens & Small Acreages." (Colorado State University Extension: 2008).

"Crickets and Grasshoppers." *Firefly Encyclopedia of Insects and Spiders*. (Firefly Books Ltd.: 2002).

Critter Repellent. "Rats in Your Yard or Garden?" http://www.critter-repellent.com/rats/rats-in-my-garden-yard.php

Cuppy, Will, and Ed Nofziger. "How to Attract the Wombat." (David R. Godine Publisher: 2002).

Davidson, Ralph H., and William F. Lyon. "Insect pests of farm, garden and orchard." (John Wiley and Sons: 1987).

Davis, David E. "The Survival of Wild Brown Rats on a Maryland Farm," *Ecology* 29, No. 4 (Oct. 1948): 437-448.

Davis, David E. "The Scarcity of Rats and the Black Death: An Ecological History." *Journal of Interdisciplinary History* 16, No. 3 (winter 1986): 455-470.

Deer Accident Statistics http://www.car-accidents.com/pages/deer-accident-statistics.html

"Deer Control in Home Gardens." Cooperative Extension Service. (West Virginia University Center for Extension and Continuing Education: March 1999).

Denmark, H. A. "The banded greenhouse thrips."

Dewey, Francis, Susan Ebeler, Doug Adams, Ann Noble, and Ulla Meyer. "Quantification of Botrytis in grape juice determined by a monoclonal antibody-based immunoassay." *American . Journal Enology and Viticulture* 51(2000): 276-282.

Discover Life. "Thrips." http://www.discoverlife.org/mp/20q?search=Thysanoptera#http://www.cals.ncsu.edu/course/ent425/compendium/thrips.html

Discover Neem Oil. 'Discover The 1001

Uses Of Neem Oil." http://www.discoverneem.com/.

Discover Life: "Black Rat." http://www.discoverlife.org/nh/tx/Vertebrata/Mammalia/Muridae/Rattus/rattus/

Discoverlife. "Norway Rat." http://www.discoverlife.org/nh/tx/Vertebrata/Mammalia/Muridae/Rattus/norvegicus/

Douthwaite, Boru, Jürgen Langewald, and Jeremy Harris. "Development and commercialization of the Green Muscle biopesticide." International Institute of Tropical Agriculture, 2001.

Dow Agrosciences Turf and Ornamental. "Spinosad." http://www.dowagro.com/turf/prod/spinosad.htm.

Drees, Bastiaan M. "A Review of 'Organic' and Other Alternative Methods for Fire Ant Control." Fire Ant Plan Fact Sheet No. 12. June 2002 revision.

Earthlife. "Aphids, Greenfly, Blackfly (Aphidoidea)". http://www.earthlife.net/insects/aphids.html

"Ecology Study of Northern Virginia Ecology: Buffalo Treehopper." Island Creek Elementary School. Fairfax County Public Schools. http://www.fcps.edu/islandcreekes/ecology/buffalo_treehopper.htmStudy

Essghaier, B. , M.L. Fardeau, J.L. Cayol, M.R. Hajlaoui, A. Boudabous, H. Jijakli and N. Sadfi-Zouaoui. "Biological control of grey mould in strawberry fruits by halophilic bacteria." *Journal of Applied Microbiology* 106 (2009): 833–846.

Evans, Edward W. "Chemical and Biological Control of Grasshoppers in Utah." (Utah State University Extension: December 1990).

Evonik Industries. "Calcium cyanamide."

http://geschichte.evonik.de/sites/geschichte/en/chemicals/inventions/skw-trostberg/calcium-cyanamide/pages/default.aspx

Excel Industries Inc. Verticel TM: Verticillium lecanii. http://www.excelind.co.in/Verticel.htm (accessed July 2010).

Ex-Pest. "Rats & Mice: Common Rat or Brown Rat." http://www.ex-pest.co.uk/rats_mice.htm

Femoralis. (O. M. Reuter) damage to ornamental plants." Proc. Fla. State hort. Soc. 89:330-331. 1976.

Flint, Mary Louise. *Pests of the garden and small farm: a grower's guide to using less pesticide*. (University of California Press: 1998).

Florida Chemical. "What is d-Limonene?" http://www.floridachemical.com/whatisd-limonene.htm

Frank, Mel and Ed Rosenthal. *The Marijuana Growers Guide*. http://www.scribd.com/doc/2601784/THE-MARIJUANA-GROWERS-GUIDE-by-Mel-Frank-Ed-Rosenthal

Gangrade, B. K. and C. J. Dominic. "Studies of the Male-Originating Pheromones Involved in Gardening Australia: Factsheet – Bordeaux Mixture." http://www.abc.net.au/gardening/stories/s1631445.htm

Gardening Zone: WHITEFLY EXTERMINATOR. http://gardeningzone.com/product_info.php?products_id=72 (retrieved July 2010).

Gardening-Guru. "Slug Traps: Death by Beer." http://www.gardening-guru.co.uk/2010/04/11/slugs-traps-death-by-beer/

Garden, M. B. "Hemp Mosaic Virus."

http://www.mobot.org/ (retrieved 2011).

Glades Crop Care. "Thrips." http://www.gladescropcare.com/tech-thrips.html

Glogoza, Phillip A., and Michael J. Weiss. "Grasshopper Biology and Management." 1997. http://www.ag.ndsu.edu/pubs/plantsci/pests/e272-1.htm

Goh, K. S., R. L. Gibson and D. R. Specker. Gray Garden Slug Deroceras reticulatum (Muller). Cornell Cooperative Extension. Field Crops Fact Sheet No. 102GFS795.00. 1988.

Golden Harvest Organic: Dealing with Whiteflies. http://www.ghorganics.com/whiteflies.html.

Golden Harvest Organics: Grasshoppers. http://www.ghorganics.com/page12.html

Golden Harvest Organics: Monterey Garden Insect Spray. http://www.ghorganics.com/MontereyGardenSpraySpinosad.html

Google docs. Biology of the Rat. http://docs.google.com/viewer?a=v&q=cache:v9YZDNSm9EQJ:education.dlam.ucla.edu/attachments/optional_reference_forms/biology_of_the_rat.pdf+biology+of+rats&hl=en&gl=us&pid=bl&srcid=ADGEEShUu4hyKnKHFsn4V6mpIO8SXbCeUZbow2Hsz90icNdt24OrRIzJqOwpC81fzl9WzK9idb0_G9k8vFrhL3PFU6fuiOZ2FQyjJBE-4QV3HMxaWDCYquTImfHFz5fmlA6vrpTlxwEE6&sig=AHIEtbTgmCGYtYeeWPDaUgl-etwxwKOVDw

"Gopher." The Columbia Encyclopedia, Sixth Edition. 2008. Encyclopedia.com. (July 26, 2010). http://www.encyclopedia.com/doc/1E1-gopher.html

Gopher Guide. Got Gophers? http://www.gopherguide.com/

Gopher Pest Control. http://www.gopherpestcontrol.com/about-gophers/

Gordon, David George. Field Guide to the Slug. Sasquatch Books. 1994.

Government of Saskatchewan. Grasshoppers. http://www.agriculture.gov.sk.ca/Default.aspx?DN=626d78ef-2444-4fae-98fc-95e05d88e759

"Grasshoppers." http://www.basic-info-4-organic-fertilizers.com/grasshopper.html

Grasshopper Facts. http://www.grasshopperfacts.net/

Great Sand Dunes National Park and Preserve: Insect Design. http://www.nps.gov/archive/grsa/resources/curriculum/elem/lesson14.htm

Green Coast Hydroponics. "How to Fight Off Root Aphids." http://docs.google.com/viewer?a=v&q=cache:mQJfK22Gfc8J:www.igrowhydro.com/InfoSheets/InfoSheet-RootAphids.pdf+root+aphids&hl=en&gl=us&pid=bl&srcid=ADGEEShaNbRz8D2GvTB93gcBc4QXldHiwtrUIFes01_btq2Ua-zw8a-P88YVjdd-UCIG3e3tHYn1QgT4b-42FeDeGc0uSN_HMb2xdu6lURk1d-RynwFAfhsOM9Y36StMbtX1ytB4F8pOZiZ&sig=AHIEtbS6GGlf-YusM0XI-9PfXRD2lFtkPEg

Green Harvest: Organic Whitefly Control. http://www.greenharvest.com.au/pestcontrol/whitefly_info.html.

GreenHome: Whitefly Control Using LadyBeetles. http://www.greenhome.com/products/pest_control/biological_pest_control/101013/ (accessed July 2010).

Greer, Lance. "Greenhouse IPM: Sustainable Whitefly Control." National Sustainable Agriculture Service:. 2000. http://

attra.ncat.org/attra-pub/gh-whitefly.html#control (accessed July 2010).

Gregner, Lance. "Deer Control Options." June 2003.

Groden, Ellie. "Using Beauveria bassiana for insect management." University of Connecticut: Integrated Pest Management. 1999. http://www.hort.uconn.edu/ipm/general/htms/bassiana.htm (accessed July 2010).

Handwerk, Brian (March 31, 2003). "Canada Province Rat-Free for 50 Years." National Geographic News (National Geographic Society). http://news.nationalgeographic.com/news/2003/03/0331_030331_rats.html. Retrieved 2007-11-30.

Hanson, Anne. "Norway Rat Behavior Repertoire" Rat behavior and biology. October 13, 2010. http://www.ratbehavior.org/norway_rat_ethogram.htm#PlayFighting

Hanson, Anne. "Rat Behavior and Biology" Rat behavior and biology. October 13, 2010. http://www.ratbehavior.org/

Hanson, Anne. "Why are rat testicles so big?" Rat behavior and biology. October 13, 2010. http://www.ratbehavior.org/testicles.htm

Hanson, Anne. "History of the Norway rat (Rattus norvegicus)" Rat behavior and biology. October 13, 2010. http://www.ratbehavior.org/history.htm

Harrod Horticultural: Whitefly. http://www.harrodhorticultural.com/HarrodSite/pages/category/category.asp?ctgry=Whitefly&cookie_test=1.

Helbig, J. Biological Control of Botrytis cinerea Pers. ex Fr. in Strawberry by Paenibacillus polymyxa (Isolate 18191). J. Phytopathology 149, 265 - 273 (2001).

Highbeam Research: Garden Snails as Escargots. http://www.highbeam.com/doc/1G1-6547821.html

Hoddle, Mark S., PhD. "The Biology and Management of Silverleaf Whitefly,Bemisia argentifolii Bellows and Perring (Homoptera: Aleyrodidae) on Greenhouse Grown Ornamentals." University of California Riverside: . http://www.biocontrol.ucr.edu/bemisia.html#verticillium (accessed July 2010).

How to Get Rid of Things. How to Get Rid of Gophers. http://www.getridofthings.com/get-rid-of-gophers.htm.

http://www.amonline.net.au/insects/insects/metamorphosis.htm

http://www.ehow.com/how_5250434_rid-caterpillars-vegetable-garden.html

http://www.for.gov.bc.ca/hfp/gypsymoth/whatisbtk.htm

http://www.howtogetridofstuff.com/pestcontrol/how-to-get-rid-of-caterpillars/

http://www.pestech.com.au/uses.htm

http://www.stretcher.com/stories/01/010903m.cfm

http://www.whatsthiscaterpillar.co.uk/america/

Hungary Starts Here: Tokaj Aszu Wine. http://hungarystartshere.com/Tokaj-Aszu-wine.

Hunter, Charles D. "Suppliers of Beneficial Organisms in North America." California Department of Pesticide Regulation. http://www.cdpr.ca.gov/docs/pestmgt/ipminov/bensuppl.htm (accessed July 2010).

Hygnstrom, Scott E., Kurt C. Vercauteren, Thomas R. Schmaderer. Biological Management(Control) Of Vertebrate Pests-Advances in the Last Quarter Century. Vertebrate Pest Conference Proceedings collection Proceedings of the Sixteenth Vertebrate Pest Conference. University of Nebraska - Lincoln (1994)

Ibiblio. H.02.01: Life Cycles & Biology of Snails & Slugs? http://www.ibiblio.org/rge/s&s.htm

Ibiblio. H.02.01: Life Cycles & Biology of Snails & Slugs? http://www.ibiblio.org/rge/s&s.htm

Illinois Natural History Survey. Section for Biodiversity. Leafhopper. http://www.inhs.illinois.edu/~dietrich/Leafhome.html

Insecta-Inspecta. How it was Transmitted. http://www.insecta-inspecta.com/fleas/bdeath/Trasmission.html

Insecta-Inspecta. The Black Death. http://www.insecta-inspecta.com/fleas/bdeath/

Insectoid.Info. Rose Leafhopper. http://www.insectoid.info/cicadas/rose-leafhopper/

Insects of West Virginia. Leafhopper. Agalliota constricta and Agalliota quadripunctata. http://www.insectsofwestvirginia.net/h/agalliota-sp.html

Insects of West Virginia. Versute Sharpshooter. Graphocephala versuta. http://www.insectsofwestvirginia.net/h/graphocephala-versuta.html

International Hunter Education Association. Guide to Wildlife Identification. Black-tailed Deer. http://homestudy.ihea.com/wildlifeID/014blacktaileddeer.htm

IPM of Midwest Landscapes. Beneficial Insects: Predatory Mites Family Phytoseiidae. Koppert Biological Systems. Lurem-TR. http://www.koppert.com/products/monitoring/products-monitoring/detail/lurem-tr/

Island Creek Elementary School

Jays, David. "Bambi and the Disney Way of Death." http://obit-mag.com/articles/bambi-and-the-disney-way-of-death

Jones, Chad C., Charles B. Halpern and Jessica Niederer. Plant Succession on Gopher Mounds in Western Cascade Meadows: Consequences for Species Diversity and Heterogeneity. Am. Midi. Nat. 159:275-286.

Journal of the Kansas Entomological Society, 49(3) (July1976), pp. 385-388. Stable URL: http://www.jstor.org/stable/25082838

Kern, Jr., William H. Southeastern Pocket Gopher. University of Florida Extension. August 2009.

Ketan, J. Solar Pasteurization of Soils for Disease. Hebrew University of Jerusaleum.

Koehler, P. G., R. M. Pereira and F. M. Oi. Ants. University of Florida IFAS Extension. http://edis.ifas.ufl.edu/ig080

Leafhopper. (2010). In Encyclopedia Britannica. Retrieved April 20, 2010, from Encyclopedia Britannica Online: http://www.britannica.com/EBchecked/topic/333835/leafhopper

"Leafhopper." Encyclopedia of Animals (2006): 1. Science Reference Center. EBSCO. Web. 18 May 2010.

Loeb, Susan D. Reproduction and population structure of pocket gophers (Thomomys bottae) from irrigated alfalfa fields. Proceedings of the Fourteenth Vertebrate Pest Conference 1990. University of Nebraska - Lincoln Year 1990.

Löns, Hermann. "A nasty animal". Supplement to the Hannover'schen Tageblatt. June 1911, Hannover.

Lubilosa Biological Control. A Serious Pest. http://www.lubilosa.org/

MacCollom, G.B. "Aphids." University of Vermont Extension Department of Plant and Soil Science. http://www.uvm.edu/pss/ppp/pubs/el60.htm

MacDonald, David W. "Pocket Gophers." The Princeton Encyclopedia of Mammals. Princeton: Princeton University Press, 2006.

MacDonald, David W. The Princeton Encyclopedia of Mammals. The Brown Reference Group. 2006.

MacGown, Joe A. "Ants (Formicidae) of the Southeastern United States." Mississippi State University. Entomological Museum. http://mississippientomologicalmuseum.org.msstate.edu//Researchtaxapages/Formicidaehome.html

Macroevolution: The Lifecycle of a Snail. http://www.macroevolution.net/life-cycle-of-a-snail.html

Marine Ornithology 37: 293–295 (2009).

Martinez, D. G. and R. L. Pienkowski. 1982. Laboratory studies on insect predators of potato leafhopper eggs, nymphs and adults. Environ. Entomol. 11: 361-362. http://www.entomology.wisc.edu/mbcn/field607.html

Matheson, Nancy. Grasshopper Management: Pest Management Technical Note. National Sustainable Agriculture Information Service. 2003.

Matthews, Ike. Full Revelations of a Professional Rat Catcher after 25 Years Experience. 1898.

Maximus L., As Observed On South Bass Island, Lake Erie." The Ohio Journal Of Science 65 (5): 298, September, 1965.

Mcguire, Betty, Theresa Pizzuto, William E. Bemis, and Lowell L. Getz. "General Ecology of a Rural Population of Norway Rats (Rattus norvegicus) Based on Intensive Live Trapping." Am. Midl. Nat. 155:221–236.

McPartland, J. M. "A Review of Cannabis Diseases." *Journal of the International Hemp Association*, 1996: 19 - 23.

McPartland, J.M. 1996. "Cannabis pests." *Journal of the International Hemp Association* 3(2): 49, 52-55. http://www.hempfood.com/iha/iha03201.html

McPartland, John. "Biological Control of Hemp Diseases and Pests." Presentation for Nova-Institute, "Online Proceedings", Bioresource Hemp 2000. Wolfsburg, Germany.

McPartland, J.M., Clarke, R.C., Watson, D.P. *Hemp Diseases and Pests*, CABI Publishing, 2000.

Meerburg BG, Singleton GR, Kijlstra A (2009). "Rodent-borne diseases and their risks for public health." Crit Rev Microbiol 35 (3). doi: 10.1080/10408410902989837. http://www.informahealthcare.com/doi/pdf/10.1080/10408410902989837.

Meyer, Wallace M. III' and Aaron B. Shiek. "Black Rat (Rattus rattus) Predation on Nonindigenous Snails in Hawaii: Complex Management Implications." Pacific Science (2009), vol. 63, no. 3:339-347.

Missouri State University: Grasshopper Plagues. http://www.lyndonirwin.com/hopper1.htm

Moorman, G. W. Penn State College of Agricultural Sciences. Tobacco Mosaic Virus. http://extension.psu.edu/plant-disease-factsheets/all-fact-sheets/tobacco-mosaic-virus-in-greenhouses

National Geographic News: Lovebirds and Love Darts: The Wild World of Mating. http://news.nationalgeographic.com/news/2004/02/0212_040213_lovebirds_2.html

National Geographic: White-tailed Deer. http://animals.nationalgeographic.com/animals/mammals/white-tailed-deer.html

National Pesticide Telecommunications Network. 1998. "Pyrethrins and Pyrethroids." Oregon State University: Corvallis, OR. http://npic.orst.edu/factsheets/pyrethrins.pdf

National Sustainable Agriculture Information Service. Greenhouse IPM: Sustainable Thrips Control. http://attra.ncat.org/attra-pub/gh-thrips.html#cultural

National Sustainable Agriculture Information Service. Sustainable Fire Ant Management: Pest Management Technical Note. http://attra.ncat.org/attra-pub/fireant.html#fire

Natural History Museum Website Exhibit on Butterfly Life Cycles. 2008. London. http://www.nhm.ac.uk/kids-only/life/life-small/butterflies/

Naturalis® L. Troy Biosciences, Inc.

NC State University: Phymatotrichopsis omnivorum. http://www.cals.ncsu.edu/course/pp728/Phymatotrichopsis/index.html#Life_Cycle

NC State University: Rhizoctonia solani. http://www.cals.ncsu.edu/course/pp728/Rhizoctonia/Rhizoctonia.html

New World Encyclopedia: Grasshopper. http://www.newworldencyclopedia.org/entry/Grasshopper

New Zealand Government: Campbell Island conservation sanctuary rat free http://www.beehive.govt.nz/node/16920

North Carolina Department of Agriculture and Consumer Services: Agronomic Division. NemaNote12: "Root-knot Nematodes on Vegetables."

North Carolina State University. Integrated Pest Management. Grasshoppers. http://ipm.ncsu.edu/ag271/small_grains/grasshoppers.html

North Carolina State University. Integrated pest management. Potato Leafhopper. http://ipm.ncsu.edu/AG271/forages/potato_leafhopper.html

Note and Queries. A Medium of Inter-Communication for Literary Men, Artists, Antiquaries and Genealogists, Etc. "The Hanover Rat." Volume Seven. January – June 1853.

ocellus. (2010). In Encyclopædia Britannica. Retrieved November 02, 2010, from Encyclopædia Britannica Online: http://www.britannica.com/EBchecked/topic/424604/ocellus

Ohio State University Extension Fact Sheet: Slugs and Their Management. http://ohioline.osu.edu/hygfact/2000/2010.html

Oklahoma State University. Entomology. Pecan Phylloxera, Phylloxera spp.http://www.ento.okstate.edu/ddd/insects/phylloxera.htm

Olkowski, William, Sheila Daar, and Helga Olkowski. 1991. Common Sense Pest Control. Newton CT: Taunton Press.

Ontario Ministry of agriculture, Food and Rural Affairs: Organic Pest Management. http://www.omafra.gov.on.ca/english/crops/hort/news/hortmatt/2003/26hrt03a7.htm

Oregon State University: Pacific Northwest Nursery IPM: Mollusks: Grey Garden Slug. http://oregonstate.edu/dept/nurspest/greygardenslug.htm

Oregon State University: Pacific Northwest Nursery IPM: Snails/Slugs: Brown Garden Snail. http://oregonstate.edu/dept/nurspest/brown_garden_snail.htm

Oregon State University: Pacific Northwest Nursery IPM: Snails/Slugs: Slugs. http://oregonstate.edu/dept/nurspest/slugs.htm

Oregon State University: Pacific Northwest Nursery IPM: Snails/Slugs: Snails. http://oregonstate.edu/dept/nurspest/snails.htm

Palaeos. Gastropoda. http://www.palaeos.com/Invertebrates/Molluscs/Gastropoda/Gastropoda.html

Pennsylvania State University: Fact Sheet: How to Pasteurize and Sterilize Containers and Tools. Pubs.cas.psu.edu/FreePubs/pdfs/xj0017.pdf

Pennsylvania State University: PA Tree Fruit Production Guide. Rose Leafhopper. http://agsci.psu.edu/tfpg/part2/insects-mites-web/rose-leafhopper

Pest Products. Sluggo Slug and Snail Bait. http://www.pestproducts.com/sluggo.htm

Pest Products: Rat Elimination. http://www.pestproducts.com/rat.htm

Pfadt, Robert E. Field Guide to Common Western Grasshoppers. http://www.sidney.ars.usda.gov/grasshopper/ID_Tools/F_Guide/grasslnd.htm

Phillips, S. W., J. S. Bale and G. M. Tatchell. "Escaping an ecological dead-end: asexualoverwintering and morph determination in the lettuceroot aphid Pemphigus bursarius L." Ecological Entomology (1999) 24, 336-344.

"Pocket gophers." Encyclopedia of Animals (January 2006): 1. Primary Search, EBSCOhost (accessed July 26, 2010).

Pyrethrins and pyrethroids. National Pesticide Information Center. Oregon State University and the US Environmental Protection Agency.

Quick Trading Co. Ed Rosenthal's GrowTips. http://www.quicktrading.com/tips1.html.

Rakitov, Roman A. 2004. "Powdering of egg nests with brochosomes and related sexual dimorphism in leafhoppers (Hemiptera: Cicadellidae)." Zoological Journal of the Linnean Society 140, no. 3: 353-381. Academic Search Premier, EBSCOhost (accessed May 18, 2010).

Ralph H. Lutts, "The Trouble with Bambi: Walt Disney's Bambi and the American Vision of Nature," Forest & Conservation History 36, no. 4 (October 1992): 160-71

Rams Horn Studio: Rat History. http://www.ramshornstudio.com/rat_history.htm

Rawlins, W. N. (1948). The Effect of Washing Frozen Mosaic Leaves on the Length of Tobacco Mosaic Virus Particles. Berkeley: University of California, Berkley.

Resource Guide for Online Disease and Insect Management. Kaolin Clay. http://www.nysaes.cornell.edu/pp/resourceguide/mfs/07kaolin.php

Resource Guide for organic pest and Disease Management. Material Fact Sheets: Beauveria bassiana. http://www.nysaes.cornell.edu/pp/resourceguide/mfs/03beauveria_bassiana.php

Resource Guide for organic pest and Disease Management. Material Fact Sheets: Neem. http://www.nysaes.cornell.edu/pp/resourceguide/mfs/08neem.php

Rocky mountain grasshopper. http://animaldiversity.ummz.umich.edu/site/accounts/information/Melanoplus_spretus.html

"Root Aphids." AskEd. http://www.cannabisculture.com/v2/articles/2967.html

"Root Aphids." Gardening Magazine. September 2007.

Royal Horticultural Society. http://apps.rhs.org.uk/advicesearch/profile.aspx?PID=515#section3

Ruisheng, An, Srinand Sreevatsan and Parwinder S Grewal. Moraxella osloensis Gene Expression in the Slug Host Deroceras reticulatum. BMC Microbiology 2008, 8:19doi:10.1186/1471-2180-8-19.

Russell, Howard and Jan Byrne. Root aphids on container grown asters. Michigan State University. Integrated Pest Management Resources.

Sadfi-Zouaoui, N., B. Essghaier, M. R. Hajlaoui, M. L. Fardeau, J. L. Cayaol, B. Ollivier and

Schalau, Jeff. Associate Agent, Agriculture & Natural Resources. University of Arizona Cooperative Extension, Yavapai County. Backyard Gardener. Managing Pocket Gophers: Part 2 - April 1, 2009. http://ag.arizona.edu/yavapai/anr/hort/byg/archive/managingpocketgopherspart2.html

Schein, Martin W. and Holmes Orgain. "A Preliminary Analysis of Garbage as Food for the Norway Rat." Am. J. Trop. Med. Hyg., 2(6), 1953, pp. 1117-1130.

Schmidt, Chris A. "Morphological and

Functional Diversity of Ant Mandibles." Tree of Life Web Project. http://tolweb.org/treehouses/?treehouse_id=2482

Scholthof, Karen-Beth Tobacco Mosaic Virus: One Hundred Years of Contributions to Virology. St. Paul: The American Phytopathological Society. 1999.

"slug." The Columbia Encyclopedia, Sixth Edition. 2008. Encyclopedia.com. (October 21, 2010). http://www.encyclopedia.com/doc/1E1-slug.html

"snail." The Columbia Encyclopedia, Sixth Edition. 2008. Encyclopedia.com. (October 21, 2010). http://www.encyclopedia.com/doc/1E1-snail.html

"Some Aphids Go Deep, Others High". Greenhouse Product News. November 2001 Volume 11, Number 12. http://www.gpnmag.com/Some-Aphids-Go-Deep-Others-High-article2797

Sparks, A.N., Jr. and T.-X. Liu. "A Key to Common Caterpillar Pests of Vegetables." Texas A&M Agricultural Communications. http://texaserc.tamu.edu

Species Page - Frankliniella tritici. http://www.entomology.ualberta.ca/searching_species_details.php?c=8&rnd=30091707&s=3454

Stange, Lionel A. (retired) and Jane E. Deisler. "Slugs of Florida". University of Florida Institute of Food and Agricultural Services. http://entomology.ifas.ufl.edu/creatures/misc/florida_slugs.htm

Stern, David L. Aphids. Current Biology Vol. 18 No 12.

Stitch, J. C. and Ed Rosenthal. "Marijuana Garden Saver: Handbook for Healthy Plants." 2008.

Stroud, Dennis C. "Dynamics of Rattus rattus and R. norvegicus in a Riparian

Habitat." Journal of Mammalogy, Vol. 63, No. 1 (Feb., 1982), pp. 151-154.

Suckow, Mark A. Steven H. Weisbroth, Craig L. Franklin. The Laboratory Rat. Academic Press. 2006

Syngenta Bioline: Whitefly control: Macro-line c (Macrolophus caliginosus). http://www.syngenta-bioline.co.uk/controldocs/html/MacrolophusCaliginosus.htm (accessed July 2010).

Taylor, Walter Penn. The deer of North America: the white-tailed, mule and black-tailed deer. Wildlife Management Institute. July 1969.

Texas A & M Ag Life Extension. Leafhopper. http://insects.tamu.edu/fieldguide/aimg88.html

Thaindian News. "Giant Rats Invade Prince Charles' Garden." http://www.thaindian.com/newsportal/world-news/giant-rats-invade-prince-charles-garden_100429533.html

The Deer of North America. Leonard Lee Rue III. 2004 (paperback edition). First Lyons Press.

The Hidden Life of Deer. Elizabeth Marshall Thomas. 2009. Harper Collins.

The idea of nature in Disney animation. David S. Whitley. Ashgate Publishing, Ltd., 2008

The IUCN Red List of Threatened Species: Rattus norviegus. http://www.iucnredlist.org/apps/redlist/details/19353/0

The IUCN Red List of Threatened Species: Rattus rattus. http://www.iucnredlist.org/apps/redlist/details/19360/0

The Living World of Molluscs. Amazing Facts About Snails. http://www.weichtiere.at/english/gastropoda/index.html

The Mammals of Texas – online edition. Mule Deer. http://www.nsrl.ttu.edu/tmot1/odochemi.htm

The Merck Veterinary Manual: Rats and Mice Overview. http://www.merckvetmanual.com/mvm/index.jsp?cfile=htm/bc/171542.htm

The Middle Ages: The Black Death: Bubonic Plague. http://www.themiddleages.net/plague.html

The University of Kentucky. College of Agriculture. Potato Leafhoppers. http://www.ca.uky.edu/ENTOMOLOGY/entfacts/ef115.asp

The Whitten Effect and Bruce Effect in Mice." Biology of Reproduction 31, 89-96 (1984).

Thomas H. Langlois. "The Conjugal Behavior Of The Introduced European Giant Garden Slug, Limax United States. Entomology Research Division. Land slugs and snails and their control. Washington, D.C. UNT Digital Library. digital.library.unt.edu/ark:/67531/metadc1506/. Accessed October 21, 2010.

Thrips of the World Checklist. Species: Frankliniella tritici (Fitch, 1855). http://anic.ento.csiro.au/worldthrips/taxon_details.asp?BiotaID=7087

trichogrammatid. (2010). In Encyclopædia Britannica. Retrieved November 06, 2010, from Encyclopædia Britannica Online: http://www.britannica.com/EBchecked/topic/604938/trichogrammatid

UMass Extension. Greenhouse Crops and Floriculture. Fact Sheets. Pest Management: A Review of Western Flower Thrips and Tospoviruses. http://www.umass.edu/umext/floriculture/fact_sheets/pest_management/wft_03.html

United States Environmental Protection Agency. Biopesticide Fact Sheet. Metarhizium anisopliae strain F52 (029056) http://www.epa.gov/oppbppd1/biopesticides/ingredients/factsheets/factsheet_029056.htm

United States Environmental Protection Agency. Pesticides: Regulating Pesticides. Metarhizium anisopliae strain F52 (029056) Biopesticide Fact Sheet. http://www.epa.gov/oppbppd1/biopesticides/ingredients/factsheets/factsheet_029056.htm

United States Environmental Protection Agency: Calcium cyanamide. http://www.epa.gov/ttn/atw/hlthef/calciumc.html

University of Alberta. Entomology Collection.

University of Arizona, Department of Plant Sciences: Whitefly Database: About Whiteflies. http://whiteflytax.biosci.arizona.edu/about_wf/index.htm

University of Arizona. Ant Information. http://insected.arizona.edu/antinfo.htm

University of California Agriculture and Natural Resources. How to Manage Pests: Pests in Gardens and Landscapes. Citrus Greenhouse Thrips. http://www.ipm.ucdavis.edu/PMG/r107301811.html

University of California Agriculture and Natural Resources. How to Manage Pests: Pests in Gardens and Landscapes. Thrips. http://www.ipm.ucdavis.edu/PMG/PESTNOTES/pn7429.html

University of California Agriculture and Natural Resources. Pests in Gardens and Landscapes: Snails and Slugs. http://www.ipm.ucdavis.edu/PMG/PESTNOTES/pn7427.html

University of California Agriculture and Natural Resources. UC IPM Online.

Grape Phylloxera. http://www.ipm.ucdavis.edu/PMG/r302300811.html

University of California Agriculture and Natural Resources: Bordeaux Mixture. http://www.ipm.ucdavis.edu/PMG/PESTNOTES/pn7481.html.

University of California Agriculture and Natural Resources: How to Manage Pests: Pests of Homes, Structures, People, and Pets. http://www.ipm.ucdavis.edu/PMG/PESTNOTES/pn74106.html

University of California Agriculture and Natural Resources: How to Manage Pests: Quick Tips for Managing Home and Landscape Pests. http://www.ipm.ucdavis.edu/QT/ratscard.html

University of California Agriculture and Natural Sciences. Key to Identifying Common Household Ants: Ant Biology and Life Cycle. http://www.ipm.ucdavis.edu/TOOLS/ANTKEY/biolmeta.html

University of California Agriculture and Natural Sciences. Pests of Homes, Structures, People

University of California, Agricultural and Natural Resources: Sooty Mold. http://www.ipm.ucdavis.edu/PMG/PESTNOTES/pn74108.html

University of California, Agricultural and Natural Resources: Whiteflies. http://www.ipm.ucdavis.edu/PMG/PESTNOTES/pn7401.html.

University of California, Agriculture and Natural Resources: Pocket Gophers. http://www.ipm.ucdavis.edu/PMG/PESTNOTES/pn7433.html

University of California, Integrated Pest Management. Green Lacewings. http://www.ipm.ucdavis.edu/PMG/NE/green_lacewing.html

University of California, Kearney Agricultural Center. FAQs About Whiteflies. http://whiteflytax.biosci.arizona.edu/about_wf/index.htm

University of California. Agricultural and Natural Resources. UC IPM Online. Seasonal development and life cycle—Leafhoppers. http://www.ipm.ucdavis.edu/PMG/GARDEN/VEGES/PESTS/LIFECYCLE/lcvegleafhopper.html

University of Florida Institute of Food and Agricultural Sciences. Featured Creatures: Greenhouse Thrips. http://entnemdept.ufl.edu/creatures/orn/thrips/greenhouse_thrips.htm#des

University of Florida, Department of Entomology and Nematology. Whitefly References. http://whiteflytax.biosci.arizona.edu/about_wf/index.htm

University of Illinois Extension: Brown Rat Management. http://ipm.illinois.edu/hyg/pests/brown_rat/

University of Kentucky Entomology. Damsel Bugs. http://www.uky.edu/Ag/CritterFiles/casefile/insects/bugs/damsel/damsel.htm

University of Kentucky Entomology. Lygaeoids: Seed Bugs and Their Kin. http://www.uky.edu/Ag/CritterFiles/casefile/insects/bugs/seedbug/seedbug.htm#geo

University of Massachusetts at Amherst. Pest Management: Managing Aphids. http://www.umass.edu/umext/floriculture/fact_sheets/pest_management/aphids.html

University of Michigan Museum of Zoology: Animal Diversity Web. Melanoplus spretus

University of Minnesota. Flower Thrips and Florida Flower Thrips. http://www.entomology.umn.edu/cues/inter/inmine/Thripg.html

University of Minnesota. IPM of Midwest Landscapes. Beneficial Insects: Predatory Mites Family Phytoseiidae.

University of Minnesota. Western flower thrips. http://www.entomology.umn.edu/cues/inter/inmine/Thripm.html

University of Minnesota: Banded Greenhouse Thrips. http://www.entomology.umn.edu/cues/inter/inmine/Thripc.html

University of Washington: Of Slug Slime, Spider Silk, and Mollusk Shells. http://www.washington.edu/research/pathbreakers/1990g.html

University of Wisconsin – Stout, Biology Department. The Rocky Mountain Locust: Extinction and the American Experience. http://www.sciencecases.org/locusts/locusts.asp.

University of Wisconsin Entomology. Know Your Friends. Damsel Bugs. http://www.entomology.wisc.edu/mbcn/kyf402.html

University of Wisconsin Entomology. Know Your Friends. Delphastus pusillus: Whitefly Predator. http://www.entomology.wisc.edu/mbcn/kyf610.html

University of Wisconsin Urban Horticulture Website. The Vermin Reports: Gophers. http://wihort.uwex.edu/pests/gophers.htm

U.S. Department of Agriculture. Agricultural Research Service. Grasshoppers: Their Biology, Identification and Management. http://www.sidney.ars.usda.gov/grasshopper/index.htm

U.S. Department of Agriculture. Agricultural Research Service. Soil Pasteurization and Inoculation with Mycorrhizal Fungi Alters Flower Production and Corm Composition of Brodiaea Laxa.

Walker, John Charles. Plant Pathology. New York: McGraw Hill Book Company, 1969.

Washington Department of Fish and Wildlife, Living with Wildlife: Pocket Gophers. http://wdfw.wa.gov/wlm/living/gophers.htm

Weiner, Irving B., Alice F. Healy, Donald K. Freedheim, Robert W. Proctor. Handbook of Psychology: Experimental Psychology. "The Rat". Pg. 52.

WineThink. The Vine's Enemy: Phylloxera Vastatrix. http://winethink.net/blog/2009/11/phylloxera/

Wiscomb, Gerald W. and Terry A. Messmer. Utah State University Cooperative Extension. Wildlife Management and Damage Series: Pocket Gophers. August 1998.

Yepsen, Roger B. 1984. Natural insect and disease control. Rodale Press.

Yu, H., J. C. Sutton. Morphological development and interactions of Clonostachys rosea and Botrytis cinerea in raspberry. Canadian Journal of Plant Pathology, 1997.

Zeyen, F. L. (2008). Tomato-Tobacco Mosaic Virus Disease. University of Minnesota.

PHOTO CREDITS

Page 10—Nitrogen deficiency: TheNew-Guy. *Page 11*—Phosphorous deficiency, top and bottom: Senseless; middle: Mynamestitch. *Page 12*—Potassium deficiency, top: MTF-Sandman, bottom: Senseless. *Page 14*—Ants tending aphids: Alexander Wild. *Page 15*—Worker ants: Alexander Wild. *Page 17*—California native ant: Alexander Wild. *Page 20*—Carpenter, Fire, Argentine ats: Alexander Wild. *Page 21*—Leafcutter ants: Alexander Wild. *Page 25*—Aphid family, © Taner Yildirim, Dreamstime.com. *Page 26*—Aphid top: © Edward Phillips, Dreamstime.com, bottom: Nature's Control. *Page 28*—Aphid convention: Ed Rosenthal. *Page 29*—Leaf damage: Ed Rosenthal. *Page 31*—Aphid: Scott Bauer, USDA Agricultural Research Service, Bugwood.org. *Page 34*—Caterpillar: Ed Rosenthal. *Page 37*—Caterpillar top: Ed Rosenthal; bottom: Digital Hippy. *Page 39*—Stem damage: Kym Kemp. *Page 40*—Caterpillar damage: Anonymous. *Page 49*—Mule deer: Walter Siegmund. *Page 51*—Native deer: Winterline Nature Trust (winterline. in). *Page 54*—Whitetailed deer: Mongo. *Page 55*—Blacktailed deer: Benjamin F. Zingg. *Page 59*—Deer fencing: Ed Rosenthal. *Page 60*—Fungus gnat: Anonymous. *Page 61*—*Bradysia* top: Whitney Cranshaw, Colorado State University, Bugwood.org. *Mycetophilidae* bottom: Jim Baker, North Carolina State University, Bugwood.org. *Page 64*—Fungus larvae: 01flat. *Page 67*—Gopher: South12th, Dreamstime.com. *Page 70*—Gopher: Dave Powell, USDA Forest Service, Bugwood.org. *Page 76*—Grasshopper: Ed Rosenthal. *Page 78*—American grasshopper: Russ Ottens, University of Georgia. *Page 81*—Two-striped grasshopper: Whitney Cranshaw, Colorado State University, Bugwood.org. Sprinkled broadwinged grasshopper: Robin McLeod. Clearwinged grasshopper: Whitney Cranshaw, Colorado State University, Bugwood.org. *Page 82*—Grasshopper laying eggs: Ed Rosenthal. *Page 83*—Leaf damage: Kym Kemp. *Page 85*—Moldy bud: Ed Rosenthal. *Page 86*—Top: Paul Bachi, University of Kentucky Research and Education Center, Bugwood.org. Middle: Rasbak via Wikimedia Commons. Bottom: Tamla Blunt, Colorado State University, Bugwood.org. *Page 87*—Moldy buds, right top: Ed Rosenthal, right bottom: Tom Flowers. *Page 90*—Leafhopper: Susan Ellis, Bugwood.org. *Page 93*—Potato leafhopper: Frank Peairs, Colorado State University. Redbanded leafhopper: Harmonia101, Dreamstime.com. *Page 98*—Scale: Whitney Cranshaw, Colorado State University. *Page 99*—Citrus mealybug: Whitney Cranshaw, Colorado State University. *Page 101*—Armored scale: Jeffrey W. Lotz, Florida Department of Agriculture and Consumer Services, Bugwood.org, via Wikimedia Commons. Hemispherical scale: Douglas Miller, United States National Collection of Scale Insects Photographs Archive. *Page 102*—Obscure mealybug: Raymond Gill, California Department of Food and Agriculture, Bugwood.org. Longtailed mealybug: Chazz Hesselein. *Page 104*—Powdery mildew: Ed Rosenthal. *Page 105*—Powdery mildew, top: Beach Stoned; middle: Cesar Calderon, USDA APHIS PPQ, Bugwood.org; bottom: Ed Rosenthal. *Page 106*—Powdery mildew: Ed Rosenthal. *Page 110*—Black rat: Mille19, Dreamstime.com. *Page 111*—Rat: Dmitry Maslov, Dreamstime.com. *Page 116*—Rat in field: Mille19, Dreamstime.com. *Page 119*—Root aphids: Tom Murray. *Page 123*—Root rot: Chemical Burn. *Page 124*—Root rot: Diggerdigzit. *Page 126*—Fusarium rot: Ed Rosenthal. *Page 130*—Slug: Connie Byrne. *Page 131*—Brown snail: Joseph Berger, Bugwood.org. *Page 132*—Banana slug: Joseph O'Brien, USDA Forest Service, Bugwood.org. *Page 136*—Leopard slug: Isselee, Dreamstime.com. *Page 139*—Spider mite webbing: Whitney Cranshaw, Colorado State University, Bugwood.org. *Page 140*—Spider mite top: Nature's Control; middle: ImageEnvision. Bottom left: Whitney Cranshaw, Colorado State University, Bugwood.org; right: Ed Rosenthal. *Page 143*—Mite webbing: Ed Rosenthal. *Page 147*—Thrip damage: Sukalo. *Page 150*— Thrip nymphs left: Andrew Derksen; right: Nature's Control. *Page 156, 158*—Tobacco mosaic virus: Oldog. *Page 160*—Verticillium wilt: David Gent, USDA Agricultural Research Service, Bugwood.org. *Page 163*—Whiteflies: Clemson University - USDA Cooperative Extension Slide Series, Bugwood.org. *Page 164*—Whiteflies top: Digital Hippy; middle: Nature's Control; bottom: Whitney Cranshaw, Colorado State University, Bugwood.org. *Page 205*—Mealybug destroyer: Sonya Broughton, Department of Agriculture & Food Western Australia, Bugwood.org. *Page 170 & 206*—Ladybugs converging: TCurtiss. *Page 207*—Ladybugs eating aphids left: Nature's Control; right: TCurtiss. Right column: Convergent lady beetle; Joseph Berger, Bugwood.org. Multicolored Asian lady beetle. Louis Tedders, USDA Agricultural Research Service, Bugwood.org. Multicolored Asian lady beetle. Scott Bauer, USDA Agricultural Research Service, Bugwood.org. *Page 208*—Rove beetle. Joseph Berger, Bugwood.org. *Page 209*—Big-eyed bug: Jack Dykinga from the ARS image gallery. [wikimedia commons]. Damsel bug nymph. Whitney Cranshaw, Colorado State University, Bugwood.org. Damsel bug adult. Alton N. Sparks, Jr., University of Georgia, Bugwood.org. *Page 210*—Minute pirate bug. Bradley Higbee, Paramount Farming, Bugwood.org. *Page 213*—*Deraeocoris brevis*. Bradley Higbee, Paramount Farming, Bugwood.org. *Page 217*—Green lacewing. Joseph Berger, Bugwood.org. Brown lacewing. Joseph Berger, Bugwood.org. *Page 218*—Lacewing nymph and green lacewing eggs. Whitney Cranshaw, Colorado State University, Bugwood.org. *Page 219*—Aphid gall midge. Predatory midge larvae in an aphid colony. Whitney Cranshaw, Colorado State University, Bugwood.org. *Page 223*—Amblyseius cucumaris. Photo: Nature's Control. *Page 224*—Spider mites. Nature's Control. *Page 225*—Predator nematode. Nature's Control. *Page 229*—Encarsia formosa. Nature's Control. *Page 231*—Trichogramma wasp. Nature's Control.

Sponsors

We would like to thank
the sponsors whose support
and participation helped
made this book possible.

INTAKE PHRESH FILTER

Intake Air & Scrubbing Filter

Available sizes.
- 4 x 6 " / 100 x 150mm
- 5 x 8 " / 125 x 200mm
- 6 x 12 " / 150 x 300mm
- 8 x 16 " / 200 x 400mm
- 10 x 16 " / 250 x 400mm
- 12 x 20 " / 300 x 500mm

- Higher air flow / lower air restriction than HEPA.
- Substantially more affordable than other filters.
- Removes 99.9% of bugs, molds, pollen and dust.
- Designed for intake filtration & "in room" scrubbing.
- Works in both directions / blow in or pull through.
- Ultra light filters using nano carbon technology.

* **Includes washable pre-filter dust sock.**

100 % RECYCLEABLE - MADE FROM 85% RECYCLED PRODUCTS

Scan to learn more about Phresh Intake Filters.
You Tube
www.phreshfilters.com
SCAN ME

VAPORIZER commercial grade

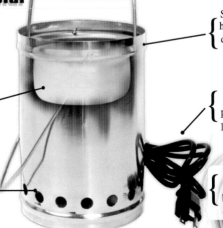

Helps prevent Bud Rot and Mould infestations.
For use in indoor gardens and greenhouses.

Stainless steel housing is very durable & will not rust

Aluminum cup maintains ideal temperature for vaporization

6 foot grounded power cord 120 / 240V.

Proprietary PTC heating element has no moving parts. It provides consistent and reliable warming.

Holes in base allow natural convection action & unrestricted air movement

Side mounted cord allows for bench mounting on flat surfaces.

Treats up to 1000 ft/2

SIERRA NATURAL SCIENCE INC.

1-877-626-5505

WWW.SIERRANATURALSCIENCE.COM

ORGANIC BOUNTEA

A COMPLETE HOLISTIC APPROACH TO PLANT NUTRITION

Brings Life to Your Plants

The Bountea Growing System will totally revitalize your soil ecology. No matter what soil type, the microbes, minerals and trace elements in Bountea ensure your soil becomes increasingly fertile with every application.

Bountea Compost Tea gives you:

Faster-growing stronger plants
Top-quality fruit and flower yields
Lush green growth throughout lifecycle
More produce with less watering
Better resistance to stress and disease

Bountea's creator, John Evans, is a master gardener with 9 World Records for Giant Vegetables, 18 Alaska State records, and over 400 first place awards for quality vegetables.

www.bountea.com

Plant Nutrition Science

The superior way to supply all the minerals hemp needs for exceptional growth in easy-to-use solutions.

Foliage-Pro 9-3-6 Plant Food

ANALYSIS

MACRONUTRIENTS	%	MICRONUTRIENTS	ppm
N	9.0	Fe	1000
P (P_2O_5)	3.0	Mn	500
K (K_2O)	6.0	B	100
Ca	2.0	Cu	500
Mg	0.5	Zn	500
S	0.05	Mo	9
		Na	500
		Cl	60
		Co	5
		Ni	1

Hemp

Cannabis Indica

SURVEY AVERAGE

MACRONUTRIENTS	%	MICRONUTRIENTS	ppm
N	5.00	Fe	99.8
P	1.07	Mn	90.4
K	3.11	B	48.7
Ca	0.74	Cu	16.2
Mg	0.36	Zn	84.3
S	0.25	Mo	2.0
		Si	602.8

Tissue analysis sample: 25 mature leaves from new growth.

© Dyna-Gro Nutrition Solutions 2010

Pro-TeKt 0-0-3
Plant Supplement

ANALYSIS

MACRONUTRIENTS

	%
K (K_2O)	3.7

MICRONUTRIENTS

	%
Si	7.8

WWW.DYNA-GRO.COM   800-DYNA-GRO

TESTING as a TOOL

Growers: Maximize Your Investment of Time & Resources

"Thank you for your dedication to providing **ACCURATE** and **DETAILED ANALYSIS**. I am happy to let you know that my new collective immediately put both strains on their "Top Shelf" list. The collective is suggesting donations of $20 per gram. This allowed me to be reimbursed at the **HIGHEST RATE PLUS A PRE-TESTED PREMIUM.** Part of my strength was presenting the Halent certified packages and ready-to-print labels for them to attach to display jars and individual sales packets.

The power of offering pre-tested medicine with professional labeling has enabled me to expand my grow potential. Professional and ethical standards are a positive in all communities and particularly illuminated in this one. Thanks again for bringing a **NEW LEVEL OF PROFESSIONAL STANDARDS** that I was able to adopt as my standard.

He told me today that patients are requesting professionally tested and certified medicines.

Looking forward to Dr. Kym's visit and learning how we can improve the quality of our medicines. We are on a monthly production schedule and look forward to seeing our results from Halent each month."

Experienced California Grower

Benefits from testing with Halent

- At least 15 Cannabinoids
- Mold & fungus testing (safety)
- At least 8 Terpenes
- Pesticides (safety)
- Verify your seedling or clone's dominant cannabinoids
  (High THC, CBD, THCV, CBC etc.)
- Grower Consultation
  (Pesticide Use, Optimized Harvest Planning, and Nutrient Use)
- Quality results can allow you an optimal return
- Crop Safety Certifications *(5 LB min.)*
- Nitrogen Packaging
- Identify medicinal value of your sample
- Affordable & confidential
- Ready to print large, medium & small labels

The Halent Scientists

PROFESSOR DONALD P. LAND   REV. DR. KYMRON DECESARE

www.halent.com

530.400.9586

Please check our website for more info from our scientists & community.

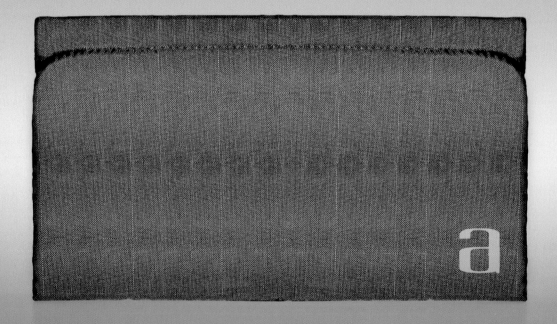

GROWONIX
TUNED FOR GROWING
™

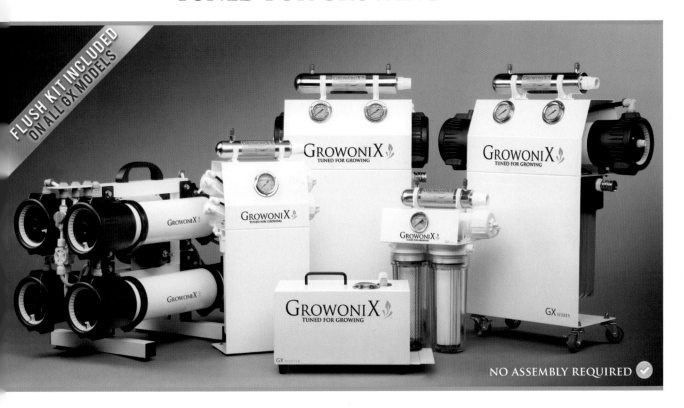

FLUSH KIT INCLUDED ON ALL GX MODELS

NO ASSEMBLY REQUIRED ✓

THE MOST EFFICIENT WATER FILTRATION SYSTEMS

FEATURING **1:1** WASTE RATIO

BUILT IN THE U.S.A.   1(888)406-2521   WWW.GROWONIX.COM